CAMBRIDGE MONOGRAPHS IN EXPERIMENTAL BIOLOGY

No. 1

THE PHYSIOLOGY OF INSECT METAMORPHOSIS

CONTENTS

LIST OF PLATES

CHAPTER I

INTRODUCTION

FROM the earliest times the insect larva has been regarded as an embryo enjoying an independent life. For Aristotle taught that the embryonic life of insects continues until the formation of the perfect insect or imago: 'the larva while it is yet in growth', he writes, 'is a soft egg.' The insect egg, according to William Harvey (1651), contains so little yolk that the embryo is forced to leave it before development is complete: it requires a larval stage in which to store up food material, until it reverts once more to the egg form or pupa. Ramdohr (1811) still retains the same idea and calls the caterpillar a 'moving, growing and feeding egg'; and so, indeed, does Viallanes (1882), who writes of 'those sorts of eggs that are called nymphs or pupae'.

Lubbock (1883) likewise traces the origin of metamorphosis to the immaturity in the condition of the insect at the time of hatching. Such an insect as the cockroach, as it develops in the egg, goes through a protopod phase, a polypod phase and an oligopod phase. According to the conception outlined by Lubbock the different groups of insects hatch from the egg at one or other of these developmental stages.

This idea, which was characterized by Pérez (1902) as 'bizarre', was later elaborated by Berlese (1913) and has been given wide currency in the writings of Imms (1925, 1937). It is often referred to as the Berlese theory. According to this 'theory' the newly hatched young of such insects as Adephaga, Staphylinidae, Meloidae, Strepsiptera or *Mantispa*, which bear a general resemblance to the primitive *Campodea*, do not leave the egg until they have reached the oligopod phase. The caterpillars of Lepidoptera, Tenthredinidae or Mecoptera, with abdominal appendages, are regarded as having hatched in the polypod phase. The larvae of other Hymenoptera and Coleoptera, with no abdominal appendages, are pictured as being between the polypod and oligopod phases, and the larvae of Diptera as being oligopod but highly modified. All these forms, which are supposed to leave the egg at an early stage of morphological development, are commonly termed 'larvae'; whereas such forms as Orthoptera, Hemiptera, Odonata, etc., which are supposed to leave the egg at a relatively advanced stage, are commonly termed 'nymphs'.

POLYMORPHISM. This approach to the problem of metamorphosis, in which the larva is seen as some kind of embryo, may have a certain value for descriptive purposes; but it is very different from the conception of insect growth that comes from a study of developmental physiology. Growth in the insect is not so very different from that in other animals. There are changes in form as growth proceeds; and since the form of the insect body is defined by an external cuticle, which remains unchanged throughout each moulting stage, it follows that any alteration in form must occur suddenly as the cuticle is shed, or must at least take place in a series of steps each more or less abrupt. This has the effect of making metamorphosis appear more catastrophic than it really is (Lubbock, 1883). But even allowing for this the changes in form are spectacular enough.

Now difference in form or polymorphism is almost universal among animals. There are differences between young and mature, between male and female, between the forms appearing in the cold or wet season and those appearing in the hot season or the dry. There are differences in caste between workers, soldiers and reproductives in ants and termites, or differences in successive or alternate generations. Metamorphosis is merely one type of polymorphism.

The polymorphic organism generally contains within it the potentialities for all its diverse forms. Sometimes the form to be developed is determined by the genetic constitution; but not uncommonly the genetic constitution is overridden by the environmental conditions experienced during growth. The 'penetrance' of genes may be influenced by temperature, or by nutrition. The genetically determined sex of Hymenoptera or Homoptera may be reversed in the course of development as the result of parasitism by Strepsiptera or dryinids. In ants and termites the potentialities to form all the female castes exist in every female egg; which castes appear is determined by the needs of their societies, acting by way of stimuli that are very incompletely understood.

Lubbock drew a distinction between animals with different terminal or mature forms (polymorphism) and animals which pass through a succession of different forms in the course of their development (polyeidism). It is doubtful whether there is anything fundamental about this distinction. The essential feature of polymorphism is multiple potentiality: one form alone is realized; the others remain latent or suppressed. The characteristic of metamorphosis is that this suppression is only temporary; so that one form replaces another during the normal life of the individual.

EMBRYOLOGY AND METAMORPHOSIS. It is obvious, as Swammerdam (1758) rightly maintained, that the latent form must exist, in some state which we do not understand, within the undeveloped organism. This applies equally to the embryonic germ in the egg; and the egg of insects has proved particularly well suited for studying the progressive determination and differentiation of this latent organism. When the egg is laid the single nucleus or germinal vesicle lies at the centre of the yolk and the egg plasma radiates from it, enclosing in its meshes the reserve substances of the yolk, to condense around the surface of the egg as the cortical plasma. One limited region of the cortical plasma, sometimes (as in *Tenebrio* (Ewest, 1937)) distinguishable visually from the remainder, and lying in the postero-ventral region of the egg, is already destined to become the embryo.

At this stage the non-cellular germ of the organism is highly plastic; it is still capable of 'regulation'. If one part is removed it is made good and a complete organism develops from the remainder. If it is divided, two complete organisms are differentiated. Then the nucleus undergoes cleavage; the daughter nuclei as they divide repeatedly move outwards and eventually reach the cortical plasma. Those nuclei which fall within the germ band zone are destined to contribute to the embryo; those which arrive elsewhere form only the extra-embryonic blastoderm.

At some stage in this process the germ band loses its capacity for complete regulation. Each part becomes committed to form some part of the embryo. The egg becomes an invisible mosaic of determined zones. This change to the mosaic state is gradual, and in many eggs some capacity for regulation persists until a late stage; but the main change can be recognized as occurring in different eggs at quite different stages of cleavage or blastoderm formation. In the dragonfly *Platycnemis*, regulation is still possible in the late blastoderm stage (Seidel, 1936); in *Sialis* determination occurs between the fourth and fifth cleavages, at the moment when the blastema is formed by peripheral migration of the cytoplasm (Du Bois, 1938); in *Bruchus* the posterior cytoplasmic regions of the egg are determined before the entrance of the cleavage nuclei, the more anterior parts shortly after the cleavage cells arrive (Brauer and Taylor, 1936); whereas in *Musca* determination of the cortical plasma is complete at the time of laying before cleavage has even begun (Reith, 1925).

In *Drosophila*, as in *Musca*, the egg at the time of laying is a mosaic egg. Local injuries effected by burning with a pencil of ultra-violet light immediately after laying cause local defects in the resulting larva. But the adult characters are unaffected; in respect to imaginal characters the egg is still capable of 'regulation'. If the egg is similarly treated 7 hours after laying or later, localized defects appear in the adult—often without any visible effect during larval development (fig. 1) (Geigy, 1931). Similar results can be obtained by puncturing the egg at different levels with a needle (Howland and Child, 1935; Howland and Sonnenblick, 1936). Likewise in *Tineola*, by burning different points in the egg with ultra-violet irradiation at different times it is possible to obtain purely larval or purely imaginal defects (Lüscher, 1944). It is clear that even at this early stage of embryonic development certain parts of the germ have already become determined to form specialized parts of the adult insect.

ORIGIN AND EVOLUTION OF METAMORPHOSIS. These observations serve to demonstrate the independent existence within the embryo of two latent organisms, larval and adult. As development proceeds the larval organism becomes differentiated, hatches and grows, while the adult organism remains in its latent, mosaic, invisible condition. This state of affairs persists until the larva is full grown; it then disappears and the adult organism becomes visibly differentiated. It is this change that is called metamorphosis. It is often regarded as a renewal of embryonic development, in that the process of transition

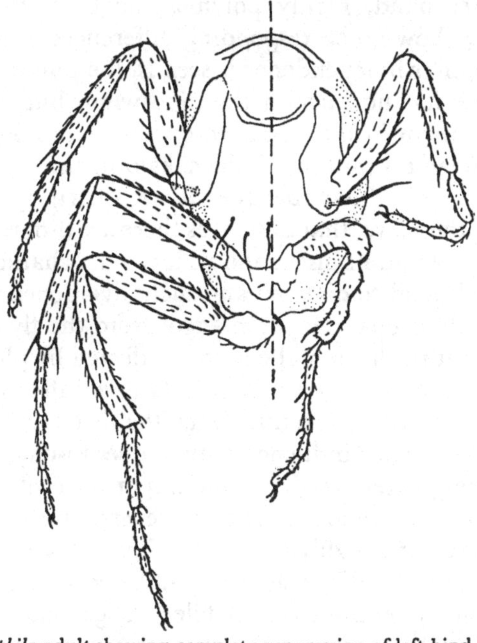

Fig. 1. *Drosophila* adult showing complete suppression of left hind-leg with deformation of first and second leg on the same side, following unilateral irradiation of the egg with ultra-violet light. Larval stages showed no visible defects. (Geigy, 1931.)

from a state of invisible determination to a state of visible differentiation is the same in both.

The important point at this stage of the argument is the independence of the alternative forms in which polymorphic organisms can exist. It is by virtue of this independence that these forms are capable of separate evolution. The same applies to the successive forms that exist in insects showing a metamorphosis. The larva is subject to the influence of its environment, and by the action of natural selection (or whatever other agencies control organic change) it may undergo an evolution of its characters which leaves the adult insect untouched. Larvae of the silkworm or of the gipsy moth *Lymantria* (Klatt, 1919) exist in many very different genetic racial forms which cannot be differentiated in the adult state. Indeed, certain species of Lepidoptera (such as *Acronycta psi* and *A. tridens*) distinguishable with great difficulty in the adult have very different larvae. On the other hand, highly polymorphic butterflies such as *Papilio polytes* show no corresponding differences in their caterpillars. A similar independence is seen in the pupa: the female of *P. dardanus* has no tails on the hind-wing, but pockets are provided for them in the pupal wing cases (Lamborn, 1914), and in the wingless mutant of the silkworm the wing cases of the pupa are formed as usual (Goldschmidt, 1927).

It follows, therefore, that as soon as there are differences in the mode of life of an adult insect and its larva the characters of the two forms will tend to diverge, and such divergence may reach the extreme that separates the maggot from the fly. Lubbock recognized that the form of the larva is affected by phylogenetic influences on the one hand and by adaptive influences on the other. Writing in 1883, at a time when the theory of recapitulation exerted a greater influence than it does to-day, he was inclined to give greater weight to the importance of phylogeny. At the present day we are inclined to reverse this emphasis: to regard the larva as a highly specialized form, closely adapted to the conditions of its life, whose resemblance to an embryo is superficial and adventitious (cf. Müller, 1869; Schindler, 1902; Snodgrass, 1953).

ORIGIN OF THE PUPA. The origin of metamorphosis is to be sought, therefore, in the divergent evolution of a polymorphic organism. The same considerations apply to the pupa in holometabolous insects. Perhaps the most obvious function of the pupa is that it bridges the morphological gap between the larva and the fully developed adult. But here again the best guide is to regard the pupa as the product of independent evolution in one form in a polymorphic organism to suit the ecological or anatomical needs of the insect in question.

Attempts have been made to find a single overriding final cause which has necessitated the development of the pupa. One of the most attractive is the view put forward by Poyarkoff (1914) and developed by Hinton (1948) that the skeletal muscles require a 'mould' in which the cuticular parts have approximately the same spatial relations as in the adult, and that a pupal mould becomes necessary when the anatomical differences are so great that the muscles required by the adult cannot be formed in the larva. But as Weber (1952) points out, the aleurodids are able to develop directly at a single moult from an extremely flattened scale-like larva to a fully winged and long-legged adult.

IMAGINAL DISCS. The classic discussions of insect metamorphosis have always centred around the more extreme examples, such as the Lepidoptera, Hymenoptera or Diptera. The anatomical study of these forms led to the discovery of the 'imaginal discs' (Lyonet, 1762), the significance of which was first appreciated by Weismann (1864), in his work on *Calliphora*, as nests of embryonic cells set aside for the formation of the adult. They represent the latent imaginal organism in visible form; and physiological discussion turned upon what factors inhibit the development of the imaginal discs in the young stages, or cause the larval tissues to disintegrate and activate the imaginal buds at metamorphosis.

It has been suggested, for example, (i) that the imaginal discs are held quiescent by the excretory products of the active larval tissues, and when these begin to degenerate the adult tissues can begin to grow (Anglas, 1901); (ii) that the larval tissues give out substances which keep the phagocytic blood cells at bay; when they age they become a prey to phagocytes and the imaginal discs develop in their place (Kowalevsky, 1887; v. Rees, 1888); (iii) that asphyxia in the larval tissues causes their degeneration (Bataillon, 1893); (iv) that the oxidase associated with melanin formation, which is at a maximum at the time of pupation, is in fact the *cause* of metamorphosis (Dewitz, 1916; cf. Agrell, 1951); (v) that the increasing size of the larval tissue cells set a physical limit to their ability to gain nourishment by diffusion (Murray and Tiegs, 1935; Tiegs, 1922); (vi) that secretions from the growing sex glands activate the imaginal discs (Pérez, 1902, 1910); (vii) that in the larva, the development of the imaginal discs is checked by secretions from the brain (Kopeč, 1924; but cf. Kopeč, 1927).

It will be observed that all these hypotheses relate to the inhibition or activation of imaginal discs. But many parts of the adult body, even in such insects as Lepidoptera or Hymenoptera which suffer a spectacular metamorphosis, are not formed from imaginal discs but are laid down by the same cells as have formed the larva. The general integument of the abdomen in adult Lepidoptera is very different from that of the caterpillar, but the same cells are concerned in its formation. In some insects the imaginal discs appear very early: in Diptera some are already evident in the late embryo (Enzmann and Haskins, 1938); in *Vanessa* the wing discs are present in the first larval stage, although those for the limbs do not appear until the end of the fourth stage (Bodenstein, 1935). But in every insect, if they are traced back far enough in development, the imaginal discs will be found to merge into the general epidermis of the larva or the embryo. There must, therefore, be a time, even in Lepidoptera or Diptera, when the same cells which are forming the visible larval structures carry latent within them the capacity to form imaginal structures.

We have already seen that this is so in the developing egg of *Drosophila* and *Tineola*. The purpose in elaborating the argument here is to emphasize that imaginal discs are not a necessary feature of metamorphosis, and that the essential problems of metamorphosis—the latency and subsequent realization of an imaginal organism—can be studied equally well in an insect in which imaginal discs do not occur.

METAMORPHOSIS IN *RHODNIUS*. It was on these grounds that the Hemipteron *Rhodnius prolixus* was chosen for the study of metamorphosis (Plate I*a–c*). *Rhodnius* has five larval stages throughout which the structural characters of the cuticle and the pigment pattern show little change. There is progressive growth and differentiation in the wing lobes, and in the later larval stages the rudiments of the genitalia begin to differentiate. But these changes are very slight in comparison with the metamorphosis which occurs when the 5th-stage larva moults to become adult.

Elaborate genital appendages are then developed: the thorax becomes elaborately shaped, with fully formed flight muscles; the wing lobes are transformed into large functional wings;

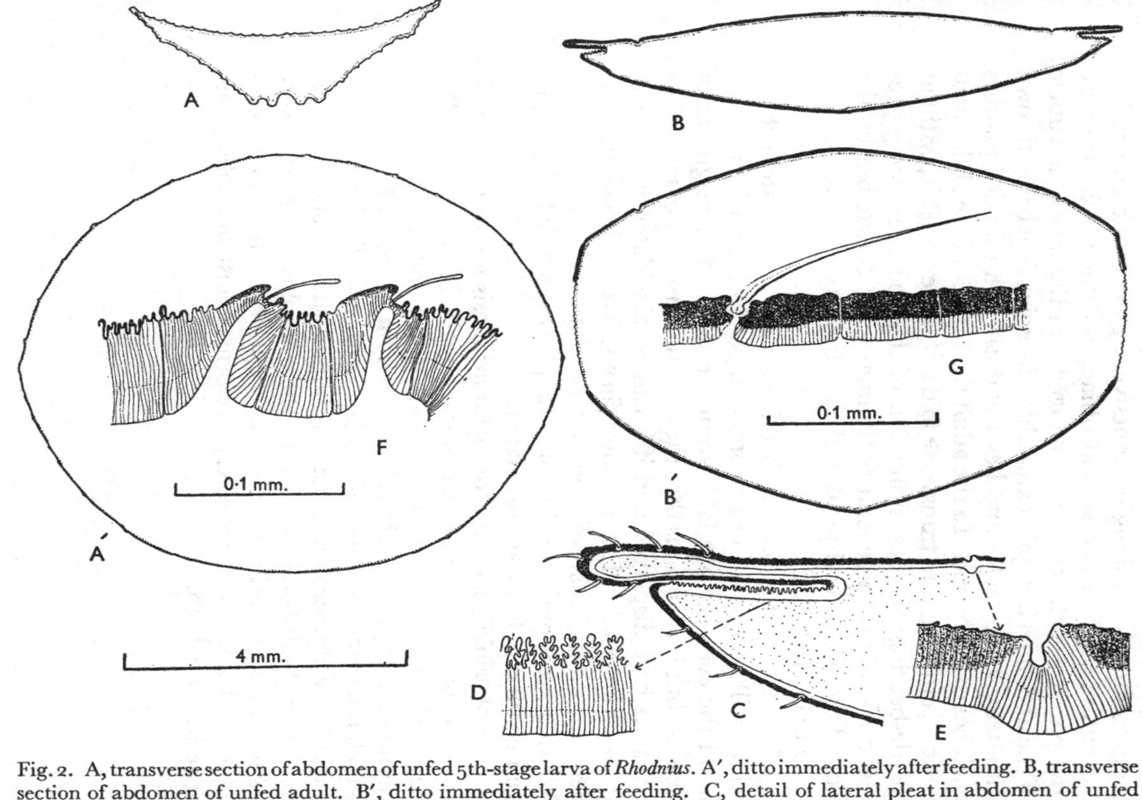

Fig. 2. A, transverse section of abdomen of unfed 5th-stage larva of *Rhodnius*. A′, ditto immediately after feeding. B, transverse section of abdomen of unfed adult. B′, ditto immediately after feeding. C, detail of lateral pleat in abdomen of unfed adult. D, detail of extensible lower wall of this pleat. E, detail of 'hinge-line' in tergites. F, longitudinal section of abdominal tergite in 5th-stage larva (cf. fig. 30A). G, longitudinal section of abdominal tergite of adult (cf. fig. 30H). (Wigglesworth, 1933.)

ocelli appear for the first time; a soft pad covered with tenent hairs to form a climbing or adhesive organ is differentiated at the end of the first and second tibiae; and the whole structure and pattern of the cuticle is changed. In the larva the cuticle of the abdomen is highly extensible, the epicuticle being thrown into deep stellate folds and beset with smooth rounded plaques each bearing a bristle. In the adult the surface of the abdominal cuticle is thrown into transverse folds; it is inextensible, without bristle-bearing plaques, distension of the abdomen being provided for, not by a general stretching of the cuticle but by the unfolding of a lateral pleat and the elasticity of a lateral strip of soft cuticle (fig. 2).

Throughout this book it is this example of metamorphosis which will occupy the centre of the stage. Full use will be made of all the evidence available from the study of other groups of insects; but it is obvious that the essential problem of metamorphosis is here displayed and if it can be elucidated in *Rhodnius* it will be equally clear in other forms. We shall not be concerned with problems of evolution, of adaptation, or of ecology. We shall treat only the physiological processes by which growth and change in form are regulated.

The physiological study of metamorphosis is concerned with the control of the emergence of the imaginal potentialities. We began this chapter with a reference to the embryological conception of metamorphosis which dates from Aristotle. The theory which interprets metamorphosis as an example of polymorphism is due to Swammerdam (1758). It implies the existence of a predetermined substrate upon which the physiological factors exert their controlling influence. The conception of development to which it leads is neither solely preformationist nor solely epigenetic. It is both.

CHAPTER 2

HISTOLOGICAL CHANGES DURING MOULTING AND METAMORPHOSIS

BEFORE discussing the mechanisms by which growth and metamorphosis in insects are controlled it will be well to describe the visible changes that take place during these processes.

GROWTH AND MOULTING IN THE INTEGUMENT. The insect cuticle is laid down by a single layer of ectodermal cells. It is this epidermis which is of prime importance in determining the form of the insect. The muscles and the motor nerves depend for their formation on the presence of the central nervous system; but the outward form of the body is controlled independently by the epidermis. If the thoracic ganglia are removed from the caterpillar of *Lymantria* a few days before pupation, the thoracic muscles of the adult fail to develop, but the adult form is normal (Kopeč, 1923; cf. Nüesch, 1952). Likewise in *Sphodromantis*, antennae and legs of normal form will regenerate after removal of the corresponding ganglia; but they contain only an amorphous granular mass without nerves or muscles (Suster, 1933). We are therefore fully justified in focusing our attention on the epidermis in our study of insect growth.

Since the epidermis is bound to the cuticle, growth can take place in it only during the periods when the epidermal cells have become detached as a preliminary to laying down a new cuticle. The process of cuticle formation and ecdysis is essentially the same during the moulting of larval stages which show little morphological change and during metamorphosis. The following account is based mainly on observations in the larva of *Rhodnius* (Hemiptera) (Wigglesworth, 1933, 1947) and in the pupa of *Tenebrio* (Coleoptera) (Wigglesworth, 1948 *b*, *d*). It has been simplified to include only what is necessary for the subsequent argument.

During the resting phase the epidermal cells may be extremely attenuated. The first sign of renewed growth consists in the 'activation' of the cells; the scattered chromatin of the nucleus becomes concentrated in a central mass and the cell body increases greatly in size, the cytoplasm becoming denser and more deeply staining. Then mitosis begins. If the new cuticle is to be much increased in size (as in the *Rhodnius* larva), or if it is to be a thick and complex structure (as in the ventral surface of the abdomen in the adult *Tenebrio*), or if many new scales or hairs are to be produced (as in the pupa of Lepidoptera), mitosis is intense. If there is to be little change in size (as in the *Rhodnius* larva induced to moult before being fed) (Wigglesworth, 1942), or if the new cuticle is to be a very thin and simple structure (as in the dorsum of the abdomen in the adult *Tenebrio*), few cell divisions occur. Even in the abdominal segments mitoses are not evenly distributed. In *Rhodnius* they begin at the margins and spread along the intersegmental membranes (Wigglesworth, 1940*b*). In the pupating larva of *Ephestia* they appear first in the thorax and spread backwards through the body (Kühn and Piepho, 1938).

CHROMATOLYSIS. The outburst of mitosis is often so exuberant that many more cells are produced than are needed to lay down the new integument. These unwanted cells break down—so that, even among the actively dividing cells are to be seen numerous deeply staining, Feulgen-positive, 'chromatin droplets' derived from the dissolution of the excess nuclei (fig. 3) (Wigglesworth, 1942, 1948*b*). Chromatolysis is particularly evident at metamorphosis, when the cells responsible for larval structures no longer required are breaking down (Poyarkoff, 1910; Wigglesworth, 1933), or when certain cells (such as the inner nuclei of the scale-forming groups in Lepidoptera (Köhler, 1932) decay almost as soon as they are produced.

It was claimed at one time (Pérez, 1910; Poyarkoff, 1910) that these chromatic droplets, described as 'boules automiques', represent the visible discharge from the nucleus of those chromatin constituents related to the specialized, larval, properties of the cells—that they constitute a visible expression of 'dedifferentiation'. But there seems no reason to doubt that they are in fact always the product of whole nuclei that have disintegrated (Wigglesworth, 1942).

NEW CUTICLE FORMATION. This fever of growth and decay among the epidermal cells is soon over. The chromatin droplets disappear, and the remaining cells and nuclei assume the orderly arrangement needed to define the form of the next instar. Where there is to be a large increase in surface area this new epithelium will be more or less folded, and such folding reaches an extreme degree where wings or other outgrowths are to be produced at metamorphosis.

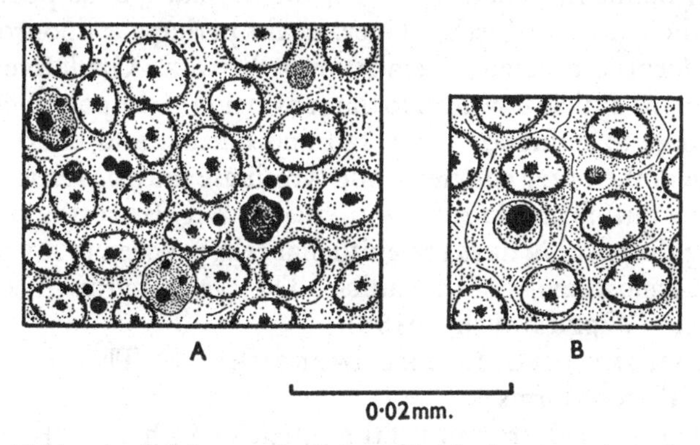

Fig. 3. Epidermis of abdomen of 5th-stage *Rhodnius* larva seen in surface view. A, 7 days after feeding; showing nuclei at an early stage of breakdown, and chromatic droplets. B, 9 days after feeding; showing a large droplet, clearly derived from an entire nucleus, invested by an adjacent cell. (Wigglesworth, 1942.)

By this time the oenocytes, specialized ectodermal cells which commonly lie between the epidermis and the basement membrane, have attained their maximum size. They now discharge their contents, which seem to be lipoprotein in nature, and this substance appears to be taken up by the epidermal cells and utilized by them in the deposition of the first layer of the cuticle —the cuticulin or lipoprotein layer of the epicuticle (Wigglesworth, 1933, 1947, 1948*b*; Wolfe, 1954).

Next, the inner layers of the cuticle, composed of chitin and protein, are laid down. Enzymes digesting chitin and protein are discharged into the moulting fluid so that the inner layers of the old cuticle are digested and reabsorbed (Plotnikov, 1904; Wigglesworth, 1933; Passonneau and Williams, 1953). Then, shortly before moulting, phenolic substances are incorporated in the outer layers of the new cuticle, and this is rendered more or less waterproof by the deposition of a thin layer of wax over its surface.

Finally, the insect swallows air to increase its bulk, ruptures the fragile remains of the old cuticle covering the head and thorax, escapes from the old skin, and by appropriate muscular effort and pressure on the body fluids expands the folded cuticular outgrowths and inflates the body to reveal its proper form. At this moment, in many insects, the dermal glands pour out upon the surface a further layer, again perhaps a lipoprotein, which forms a protective cement over the wax. Oxidative processes in the cuticle now lead to the conversion of the polyphenolic substances to quinones which tan and harden and darken the protein of the outer cuticle; while tyrosinase may be responsible for melanin formation (Pryor, 1940; Wigglesworth, 1949). During the ensuing days the epithelial cells proceed to lay down the innermost layers of the cuticle. The cycle of growth is then complete (Wigglesworth, 1948*d*).

METAMORPHOSIS IN THE INTEGUMENT. This may take many different forms.

(i) The general structure of the cuticle may change. In caterpillars the cuticle is mostly soft and extensible with no hard exocuticle; in the pupa a rigid exocuticle composes half the total thickness (Kühn and Piepho, 1938). In the *Rhodnius* larva the cuticle of the abdomen is beset with little bristle-bearing plaques with extensible regions between them in which the epicuticle is thrown into deep stellate folds; in the adult the bristles have mostly disappeared, at least from the tergites, the surface folding is mostly transverse and a hard exocuticle renders the integument inextensible (fig. 2F, G). In the pupa of *Tenebrio* the dorsal and ventral cuticles of the abdomen are alike; but in the adult beetle the tergites become delicate membranes no more than 4μ in thickness, the sternites form thick and horny plates (36μ) rich in peculiar dermal glands and numerous small hairs (Wigglesworth, 1948*b*).

(ii) There may be striking changes in the form of the articulated hairs, and these take place, although the hairs are laid down by the same trichogen and tormogen cells persisting from one instar to the next. This is very evident in *Rhodnius* (Wigglesworth, 1933). In the larva of *Ephestia* there may be substantial changes in the form of the setae in successive larval instars and still more striking changes at pupation (Krumiņš, 1952).

(iii) There may be an entirely new vestiture of hair or scales. This is particularly striking in the change from the pupa to the adult in Lepidoptera, when great numbers of new scale-forming cells are differentiated from among the epidermal cells which laid down the uniform smooth cuticle of the pupa (Stossberg, 1938).

(iv) Finally, there may be large changes in general form which result from enormously enhanced local growth. The details of this process vary greatly, but the essentials are the same whether the structures in question are elaborate genital appendages, or thoracic excrescences, such as develop in the adult *Rhodnius*, or the wings themselves. The process of intensive local growth that results in wing formation has been studied histologically in most detail in the pupa-adult transformation of Lepidoptera (Köhler, 1932; Stossberg, 1938).

It very often happens that the adult insect is more conspicuously differentiated. But this is not always so. The elaborate abdominal appendages of caterpillars, or the abdominal gill plates in the larvae of mayflies, or the feeding brushes of mosquito larvae, will bear comparison with the wings or the specialized feeding apparatus of adult insects. What we witness in metamorphosis is change in form in conformity with a changed way of life and *not* the progressive differentiation of a comparatively undifferentiated organism.

IMAGINAL DISCS AND CELL SIZE. In all the examples of metamorphosis so far considered in this chapter the transformations observed have been carried out by the same cells as were responsible for building the larva. But in the larvae of endopterygote insects certain groups of epidermal cells become set aside for the formation of the pupal and imaginal structures. Early in life these cells may form part of the larval epidermis engaged in laying down the larval cuticle. But later they become invaginated as little epithelial pockets. Later still there is an evagination of epithelium outwards into the lumen of this pocket, and it is this evagination which forms the germ of the wing or appendage or whatever it may be (fig. 4).

These imaginal discs, as we have seen (p. 8), arise at very different times in the growth of different insects. In the limbs of Lepidoptera, for example, they are not visibly distinguishable from the general epidermis until the end of the fourth instar (Bodenstein, 1935). They attain their most specialized development in the higher Diptera, and we must consider rather more fully the histology of metamorphosis in this group of insects.

In such insects as *Musca*, *Calliphora* or *Drosophila* the imaginal discs become invaginated from the ectoderm towards the end

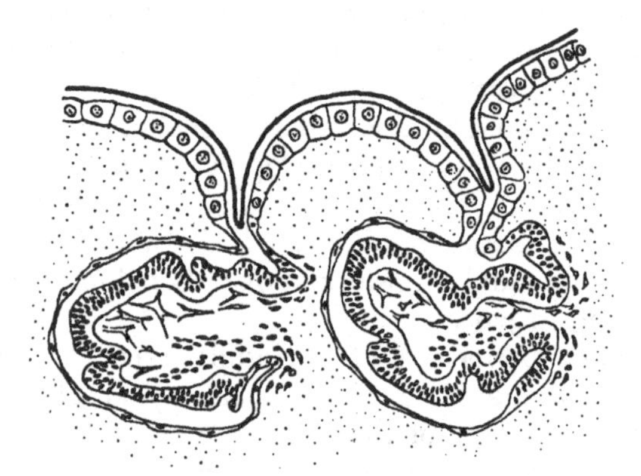

Fig. 4. Longitudinal section through the imaginal discs of the wings in full-grown larva of the ant *Formica*. (Pérez, 1902.)

of embryonic life or at a very early larval stage. At this time all the epidermal cells are quite small; but thereafter, as was pointed out by Pérez (1910) in *Calliphora*, those cells which form purely larval structures—the epidermis, fat body, salivary glands, etc.—cease to multiply, and grow only by increase in cell size. The cells of the imaginal discs, on the other hand, remain small and the discs grow by the multiplication of their cells.

Attempts have been made to generalize on this matter of cell size and metamorphosis. In the chalcid *Mormoniella* (*Nasonia*) there is likewise no cell division in many of the larval organs; the large cells of these break down at metamorphosis, and Tiegs (1922) even suggested that cell size, by setting a limit to the nutrition of the cell by diffusion, may thereby set a limit to larval growth and thus constitute the controlling factor in metamorphosis. In the silkworm, most of the tissues combine increase in cell size and cell number, and they persist through metamorphosis—with the exception of the large-celled silk glands which degenerate (Trager, 1937). In the larva of the mosquito *Aedes* the cells of epidermis, fat body and thoracic ganglia grow by cell multiplication and persist through the pupa to the adult; while the trichogen cells, anal gills, fore-gut, salivary glands, etc., grow by increase in cell size and are destroyed (Trager, 1937). But there are plenty of exceptions to this generalization; for in this same insect the Malpighian tubes, the heart and pericardial cells all grow by cell hypertrophy and yet persist into adult life; cell size is certainly not the main factor determining the fate of insect cells during metamorphosis (Trager, 1937).

Some of the increase in cell size during larval growth is functional and temporary; in the caterpillar *Ptychopoda* there is a doubling of nuclear volume in the epidermis before each mitosis but no polyploidy (until the large polyploid scale-forming cells appear in the pupa). On the other hand, the fat body cells do become polyploid (Risler, 1950). In the larva of *Calliphora* the enormous increase in cell size is attributed to endomitosis leading to polyploidy (Wagner, 1951). But there is no obligatory breakdown of polyploid cells at metamorphosis. In the larva of *Culex* the cells of the hind-gut become progressively more highly polyploid at each larval moult. At metamorphosis, however, a series of reduction divisions take place; and the daughter cells survive in the adult, the normal diploid condition having been restored (Berger, 1938).

The imaginal discs of *Calliphora* and other higher Diptera become virtually detached from the larval epidermis, being connected only by a tenuous thread, the vestigial neck of the invagination. But in addition to these compact embryonic buds, there are numerous independent histoblasts or small groups of cells scattered throughout the body, notably beneath the larval epidermis. At the time of pupation (which takes place beneath the hardened larval skin or puparium) the pupal cuticle covering the head and eyes and limbs, which are evaginated at this time, is the product of the cells of the imaginal discs. But the cuticle covering the abdomen is laid down by the larval cells; not until after pupation are these replaced by a continuous layer of imaginal cells arising by proliferation from the groups of histoblasts (Wolfe, 1952).

Immediately after pupation in these flies histolysis of the purely larval cells begins. This is a process identical in nature with the chromatolysis already described in the epidermis of *Rhodnius* or *Tenebrio*. But since many more cells have been specialized for purely larval purposes this histolytic process is correspondingly exaggerated. The tissues break down in sequence; and the last to go is the fat body, larval cells of which persist for some time even in the adult, although the adult fat body is already fully formed (Pérez, 1910).

Thus, in all insects, metamorphosis consists of three histological processes: total destruction of the most specialized larval structures; completely new constructions of the most specialized organs of the adult from embryonic histoblasts; and reconstruction *in situ* of the less specialized and more plastic organs and cells. These three processes are superimposed and intermingled in varying degrees in different insects (Pérez, 1910).

METAMORPHOSIS IN INTERNAL ORGANS: PHAGOCYTOSIS. There is the same degree of variation in the metamorphosis of the internal organs. These may pass with little change from the larva to the adult or they may be completely replaced from embryonic cells. For example, the midgut of *Calliphora* is re-formed from 'imaginal rings' at the points of union with fore-gut or hind-gut (Pérez, 1910); in the ant *Formica* it is rebuilt from histoblasts or regenerative cells along its length (Pérez, 1902); in *Hepialus* it persists with little change (Henson, 1946). It is worth noting that in some insects (termites, *Blattella*, etc. (Weyer, 1935, 1936)) the mid-gut epithelium is replaced at each moult. Such renewal is therefore not peculiar to metamorphosis. Similar degrees of variation are to be seen in the metamorphosis of the Malpighian tubes, salivary glands, and skeletal muscles (Henson, 1946).

In the removal of the histolysing tissues the phagocytic blood cells may play a considerable part, but it is now generally agreed that they are concerned only in removing cells which are already dead and in process of autolysis—although such cells may show little histological change (Pérez, 1910; Tiegs, 1922). In the blowfly *Calliphora* haemocytes stuffed with granules of disintegrating tissue, the 'Körnchenkugeln' of Weismann, are abundant in the blood of the pupa (Pérez, 1910); whereas in the allied blowfly *Lucilia*, autolysis and fragmentation of the muscles occur spontaneously; the haemocytes appear late on the scene and are concerned only in the ultimate digestion of the muscle fragments (Evans, 1936). In the ant *Formica*, salivary glands, Malpighian tubes, fat body cells and many muscles are phagocytosed (Pérez, 1902); in the chalcid *Mormoniella*, many of the tissues dissolve without the intervention of haemocytes, but these fall upon any undissolved fragments present in the blood, and some tissues are attacked when they show little visible change; here autolysis and phagocytosis take place side by side (Tiegs, 1922); while in the honey bee there is no evidence of phagocytosis in any organ, whether muscle, fat body, silk glands or gut wall (Oertel, 1930).

GROWTH RATIOS. Certain laws appear to govern the change in size or proportions of the body that take place at moulting and metamorphosis. The earliest of these laws was propounded by Dyar (1890), who pointed out that the width of the head in caterpillars increases by a fairly constant ratio at each moult, a ratio which varies from species to species but is usually about 1·4. This rule has been found to apply pretty generally to many of the parts of insect larvae. There is, in fact, a geometrical increase at each moult, so that if the logarithm of the linear measurement of the part is plotted against the number of the instar a straight line is obtained (Teissier, 1936).

This 'law' is merely a description of what normally occurs in many species. It is by no means a necessary consequence of the moulting process. If the insect is ill fed its growth is reduced and the 'progression factor' or ratio of increase falls with age (Beck, 1950). In certain insects the value is different at different moults: in the larva of *Popillia* the factor for the length of the head falls progressively (1·89, 1·65, 1·0) at successive moults (Abercrombie, 1936); and many other exceptions occur (Gaines and Campbell, 1935).

In *Sphodromantis* the linear growth ratio is in the region of 1·25 and the ratio for weight in the region of 2·0. Przibram and Megušar (1912) therefore concluded that during each moulting stage every cell in the body divides once and grows to its original size. But we have seen that the histology of growth and moulting is something very different from this. It is true that the number of nuclei in the epidermis of *Sphodromantis* (Sztern, 1914) or *Rhodnius* (Wigglesworth, 1940*a*) is approximately the same in an area of given size at all stages of growth. But that does not necessarily mean that it is the division of cells that has determined the changes in surface area that occur. Indeed, the number of cells present is often a secondary consequence of the size of the body as determined by other means. If the abdomen of *Rhodnius* is stretched by blocking the anus immediately after feeding, the abdomen of the next instar is abnormally large; and there is a compensatory increase in the number of epidermal cells to cover the increased area (Wigglesworth, 1940*a*). Moreover, in Muscidae and other flies, the rigid parts of which grow in fair agreement with the numerical rule (Alpatov, 1929), the larval cells do not divide after the embryonic period but grow in size only.

In hemimetabolous insects such as *Pediculus* there is an abrupt change in the progression factor at the final moult (Buxton, 1938). This, of course, is a necessary consequence of any metamorphosis, a process which by definition is characterized by a more or less striking change in form. It follows, also, that whereas growth throughout the larval stages tends to be 'harmonic', the dimensions of the parts increasing at approximately the same rates as one another so that the proportions of the body are little changed, metamorphosis is characterized by the exaggerated growth of particular parts. This exaggerated, disproportionate, or 'disharmonic' growth tends to follow the well-known allometric law (Huxley, 1932); that is, the parts grow at rates peculiar to themselves, higher or lower than the growth rates of the body as a whole.

This relation is expressed by the formula $y = bx^k$, where x is the dimension of the whole, y the dimension of the part, k the 'growth coefficient' and b another constant. That is, the logarithm of the dimension of the part is proportional to the logarithm of the dimension of the whole, so that when these measurements are plotted on a double logarithmic grid a straight line is obtained. Thus, looked at from the point of view of growth ratios, metamorphosis is seen as an abrupt change in the distribution of growth activity in the different parts of the body, an exaggeration of allometric growth (Novák, 1951*a*, *b*).

CHAPTER 3

THE PHYSIOLOGY OF GROWTH AND MOULTING

THE epidermis, as we have seen, is the chief tissue responsible for growth and form in the insect. At each instar the visible form of the body is defined by the cuticle which is laid down by and is indeed an integral part of the epidermis. In the softer parts of the integument the cuticle can unfold and stretch, but in the more rigid skeletal structures, such as the head and the appendages, growth is impossible unless the cuticle is shed. It follows, therefore, that growth and moulting are intimately associated; indeed, the study of the physiological control of growth resolves itself into a study of the control of moulting.

We have seen that in holometabolous insects certain structures which show an excessive degree of growth at the time of metamorphosis become invaginated from the epidermis and separated from it so that they are relieved of all responsibility for cuticle formation during larval life, and are thus able to grow more or less continuously. We shall consider later the growth of these imaginal discs; for the moment we may neglect them and consider merely the phenomenon of growth and moulting as such.

HUMORAL CONTROL OF MOULTING. The fact that moulting affects all parts of the body at the same time, early suggested some kind of central control. The first experimental demonstration of this was given by Kopeč, 1917, 1922). Kopeč showed that the last-stage larva of *Lymantria* ligated with a thread before a certain critical stage of growth will undergo pupation in front of the ligature but remain a permanent larva in the posterior half. Section of the nerve cord has not this effect; whence he concluded that a hormone is concerned in inducing the pupal moult.

This experiment has been repeated in caterpillars and other insects, always with the same result. *Calliphora* larvae ligatured more than 16 hours before pupation form the puparium in the anterior half only; but they can be induced to do so in the posterior half also if the blood of larvae about to pupate is injected (Fraenkel, 1935*a*). The larva of *Rhodnius* or *Cimex* feeds only once in each instar; and if it is decapitated within a day or two

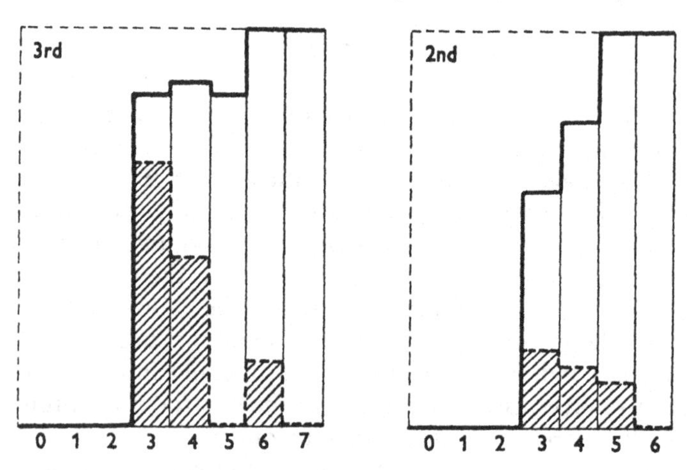

Fig. 5. Charts showing the proportion of *Rhodnius* larvae in 3rd and 2nd stage which moulted out of batches decapitated at different times after feeding. Figures on the base-line represent days after feeding. The uppermost level represents 100 per cent moulting. The broken line (above shaded area) shows the proportion of insects developing adult characters. (Wigglesworth, 1934.)

after feeding it never moults. But there is again a critical period, when a sufficient amount of the moulting hormone has been secreted into the blood, and thereafter moulting is not prevented by decapitation (fig. 5) (Wigglesworth, 1934). Here again, if a larva decapitated soon after feeding is joined by means of a capillary tube to a larva decapitated after the critical period, (Plate II*a*) so that the blood of the two insects is mixed, it is induced to moult (Wigglesworth, 1934, 1936).

In these so-called 'parabiosis' experiments, even when the insects are joined by a glass or plastic tube, the epidermis shows a remarkable capacity for spreading and uniting the one insect with the other (Williams, 1948*b*), and even when these belong to different families (as in the case of *Rhodnius* and *Cimex* (fig. 6) (Wigglesworth, 1936)). The possibility therefore remained that the controlling influence might spread from cell to cell and not be carried in the circulating blood. The same criticism applies to experiments in which the limbs or complex spines of caterpillars were transplanted from one insect to another and shown to moult at the same time and to undergo the same number of moults as their new host (Bodenstein, 1933).

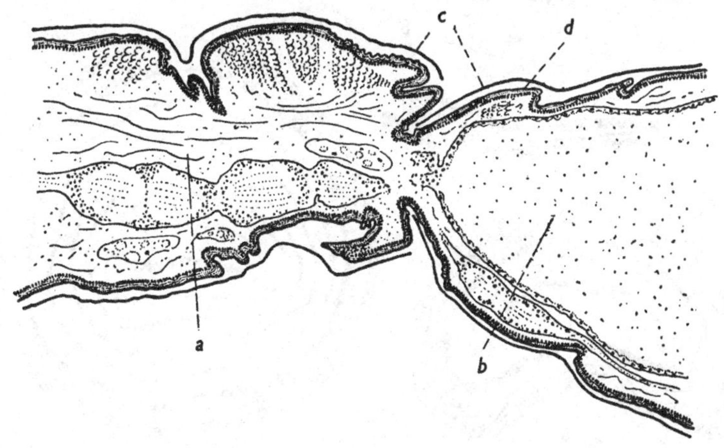

Fig. 6. Longitudinal section through the point of junction between a decapitated 3rd-stage *Rhodnius* larva and a 5th-stage *Cimex* larva transected behind the prothorax, showing the continuous epidermis and cuticle. *a*, *Rhodnius*; *b*, *Cimex*; *c*, old cuticle; *d*, new cuticle. (Wigglesworth, 1936.)

A different method of approach was furnished by the study of hybrid Lepidoptera. It has long been known that in such hybrids growth is more liable to be affected in the female than in the male. If the sphingids *Celerio gallii* ♂ and *C. euphorbiae* ♀ are crossed, the female pupae fail to produce imagines, though they may survive for several years; the males complete their normal development. If the wing germs or ovaries of the hybrid females are transplanted into the hybrid males they develop normally. It was therefore suggested by Bytinski-Salz (1933) that it is the growth hormones that are lacking in the female pupae. And it has in fact been shown by Meyer (1953) that it is possible to induce development in the female hybrid pupae by withdrawing a large quantity of their blood and replacing it by the blood of male pupae in which development has already begun.

If a small cylindrical segment of the head or of a limb in *Rhodnius* is implanted into the abdomen, the epidermal cells grow outwards, spread over the outer surface of the cuticle and

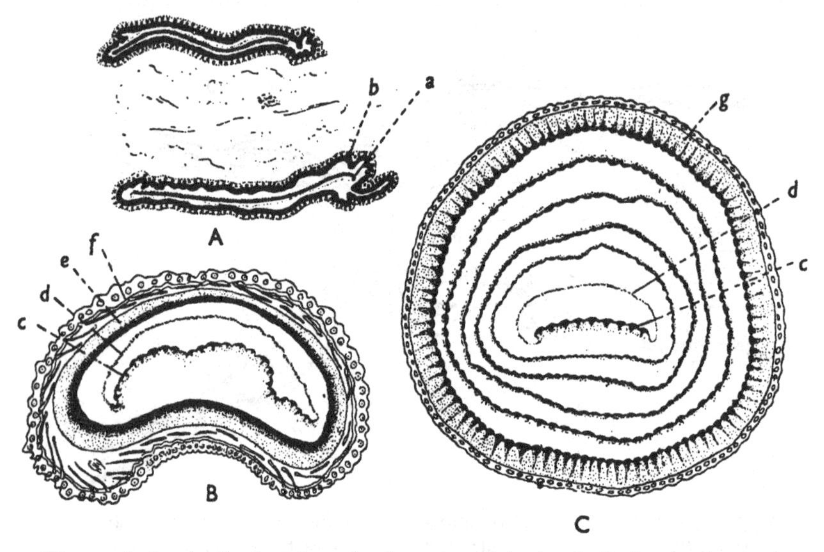

Fig. 7. A, longitudinal section of a fragment of the head of 4th-stage larva of *Rhodnius* after implantation into a moulting 5th-stage larva. *a*, original cuticle; *b*, new cuticle and epidermis which has grown around to form a closed capsule (Wigglesworth, 1936). B, schema of an implanted fragment of the integument of *Galleria* larva after metamorphosis of the host. *c*, implanted larval cuticle; *d*, thin cuticle shed from the epidermis that has grown round; *e*, pupal cuticle; *f*, imaginal cuticle with scales (after Kühn and Piepho, 1940). C, schema of implanted fragment of integument of full-grown larva of *Galleria* caused to make five extra moults by transplanting to a young larva. *g*, the latest formed larval cuticle with epidermis (after Piepho, 1943).

unite with the cells from the other end to form a closed capsule which then proceeds to undergo a series of moults simultaneously with the host (fig. 7A) (Wigglesworth, 1936).

Implants of this kind have been used extensively in Lepidoptera (*Galleria*, *Ptychoptera*, etc.) in order to study the hormonal control of moulting and metamorphosis (Piepho, 1938*a*, *b*). It has been shown, for example, that implants made into 'permanent' larvae (resulting from ligature before the critical period) heal round in the usual way but do not moult. Whereas in the normal larva they moult simultaneously with their host until metamorphosis is complete (fig. 7B) (Piepho, 1938*a*). The same occurs when the integument of *Ptychoptera* is implanted in *Galleria* (Piepho, 1938*c*). And if a fragment of integument from a full-grown larva is implanted into a young larva it will make a series of extra moults; it may then be removed and reimplanted into a young larva and the process thus continued, probably indefinitely (fig. 7C) (Piepho, 1938*c*). The removal and reimplantation of these fragments at different intervals confirms the conclusions reached from the ligature of whole insects, that during the critical period they become independent of further supplies of the hormone (Piepho, 1939*b*).

THE BRAIN AND MOULTING. It was early suggested that the initial stimulus to moulting comes from the brain. Kopeč (1917, 1922) showed that the larva of *Lymantria* is prevented from moulting to the pupa if the brain is removed before the critical period of the final instar; whereas ligature of the nerve cord, or excision of the suboesophageal ganglion, has not this effect. He therefore suggested that the brain was the source of the hormone necessary for moulting.

Plagge (1938) confirmed these results on the larva of *Deilephila* and succeeded in getting many of the larvae to pupate after reimplantation of the brain. Kühn and Piepho (1936) repeated these experiments on *Ephestia* but obtained very few positive results (4 out of 110 implantations)—probably, as we now realize, because the implanted brains were not taken from larvae at just the right stage of secretion. Similarly in the sawfly *Trypoxylon*, pupation is prevented by extirpation of the brain, but can be induced in larvae after ligature round the neck by the injection of blood from pupating larvae or the implantation of the brain into the abdomen (Schmieder, 1942). Removal of the brain in *Bombyx mori* larvae during the final instar shows that there is a critical period for the pupal moult at the end of the phase of obligatory feeding. Then there is a second critical period, for the imaginal moult, at the time when spinning of the cocoon begins (Bounhiol, 1952*a*, *b*).

In *Rhodnius* it was at first suggested (without good experimental evidence) that the hormone causing moulting comes from the corpus allatum (Wigglesworth, 1934). But this was disproved in Lepidoptera, for caterpillars continue to moult after the corpus allatum has been removed (Bounhiol, 1938; Plagge, 1938). Histological examination of the brain in *Rhodnius* revealed the presence of large neurosecretory cells, stuffed with fuchsinophil droplets, confined to the 'pars intercerebralis' of the protocerebrum (Hanström, 1938). If this region of the brain is excised from *Rhodnius* larvae one week after feeding, that is, just around the 'critical period', and implanted into the abdomen of larvae decapitated soon after feeding, these are

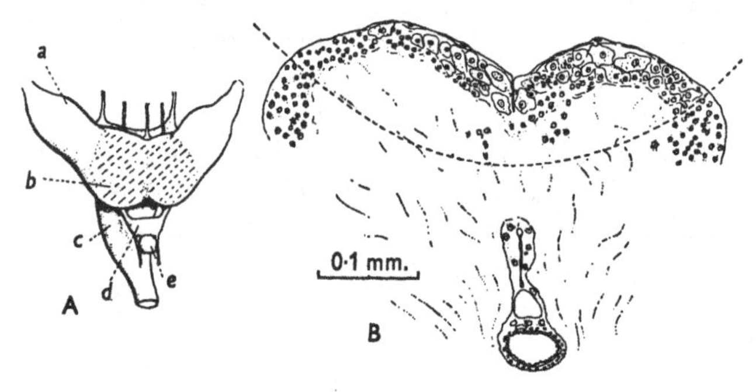

Fig. 8. A, brain of 5th-stage *Rhodnius* larva. *a*, optic lobes; *b*, protocerebrum; *c*, suboesophageal ganglion; *d*, corpus cardiacum; *e*, corpus allatum. The shaded area of the protocerebrum shows the approximate extent of the region excised and implanted. B, vertical section through the posterior part of the protocerebrum showing neurosecretory cells in the central region. The broken line marks off approximately the region implanted. (Wigglesworth, 1940*b*.)

caused to moult. No other part of the brain has this effect; nor has the suboesophageal ganglion or the corpus allatum (fig. 8) (Wigglesworth, 1940*b*).

THE BRAIN AND NATURAL ARREST OF GROWTH. If the larva of *Rhodnius* receives too small a meal of blood, it will not moult; indeed, it may be given a succession of small meals so that the stomach always contains plenty of food and yet moulting does not occur. In this insect it seems to be the stretching of the abdomen, which provides the nervous stimulus to the brain, that is needed to bring about the necessary activity in the secretory cells. Likewise, as Kopeč showed, caterpillars deprived of the brain can survive as 'Dauer-larvae' for many weeks; and larvae of *Rhodnius* decapitated in the 5th stage within a day after feeding have remained alive for more than a year. Such insects could be described as being in a state of diapause.

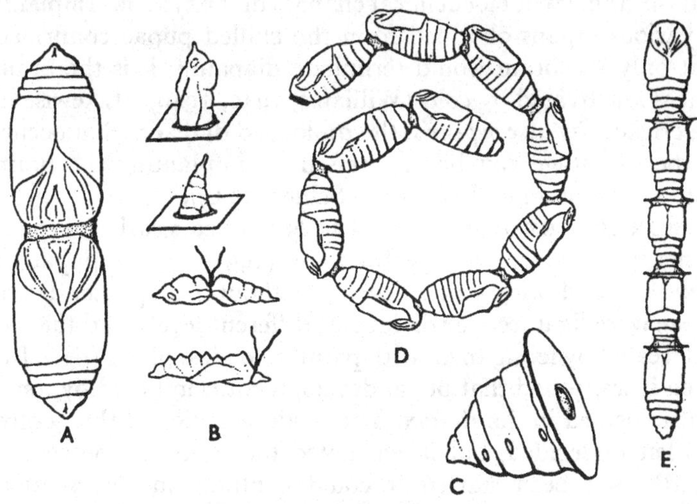

Fig. 9. A, one of the experiments of Crampton (1899) to show simultaneous development of *Platysamia* pupae joined with paraffin. B, experiments of Hachlow (1931) to demonstrate the presence of a thoracic centre causing development in pupae of *Aporia* and *Vanessa*. C–E, experiments of Williams (1952) on *Platysamia* pupae. C, isolated abdomen caused to develop by implantation of prothoracic gland and chilled brain. D, chain of ten brainless diapausing pupae induced to develop by implantation of a chilled brain into the leading pupa. E, four pupal abdomens in diapause joined to a chilled pupa with brain and prothoracic gland; only the anterior-most abdomens develop.

On the basis of these observations it was suggested that the natural arrest of growth that occurs in insects in diapause may be due to a temporary failure in the secretion of growth hormone (Wigglesworth, 1934). This has been found to be the case in the overwintering pupae of the large silk moths *Platysamia* and *Telea*. These enter diapause immediately after pupation and, as in many other insects, diapause is brought to an end by exposure at 3–5° C. for one or two months. Williams (1946), using the method of Crampton (1899) (fig. 9A), joined 'diapause pupae' in parabiosis with 'chilled pupae' but could obtain no evidence of the presence of an inhibitor of development in the diapausing pupae; diapause appeared to be due simply to the *absence* of growth factors.

These diapause pupae and chilled pupae have provided admirable material for studying the hormonal control of growth and the associated biochemical changes in the system. Implants of various organs obtained from the chilled pupae confirmed that only the brain would terminate diapause; it is the brain that is reactivated by cold (Williams, 1942, 1946). Likewise in the cricket *Gryllus campestris*, the prolonged diapause that occurs in the 9th stage can be suppressed by implanting the brain from larvae of an earlier instar (Sellier, 1949).

THORACIC GROWTH CENTRE. Many years earlier, experiments were carried out by Hachlow (1931) on the pupae of *Vanessa* and *Aporia*, which develop without a diapause. These pupae were ligatured, or divided at different levels, and the cut ends sealed by fixing them with paraffin to glass plates (fig. 9B). Then it was found that pupal development is initiated by some centre located in the thorax. The mode of action of this centre was left undecided, but it was noted that once the process of growth had been started it could continue in the isolated abdomen. These results were confirmed in *Phryganidia* (Lepidoptera) by Bodenstein (1938).

This thoracic centre was identified by Fukuda (1940*a*, 1944) with the 'prothoracic gland', a diffuse organ, made up of bead-like strings of large cells applied to the trachea of the first thoracic segment of caterpillars (fig. 10) (Lee, 1948). This gland was figured by Lyonet (1762) in the larva of *Cossus* under the name of the 'granulated vessels'; it was observed by Verson and Bisson (1891) in the silkworm and called the 'hypostigmatic gland', and its origin in the embryo as an ingrowth from the second maxillary segment was described by Toyama (1902).

If the larva of the silkworm is ligatured behind the prothoracic gland, only the anterior half pupates. But the posterior part can be induced to pupate if the prothoracic gland is implanted into the abdomen. At the critical period the prothoracic gland releases an active principle into the blood (Fukuda, 1940*a*). This same gland controls the moulting of the larva (Fukuda, 1940*b*) and the development of the pupa. The isolated abdomen of the pupa will resume its development if it is connected to the anterior part of the body by a capillary tube, or if the prothoracic gland removed from another pupa or from a larva is implanted in it (Fukuda, 1944). And the larval prothoracic gland will replace the pupal gland in terminating diapause in *Platysamia* (Williams, 1952).

ACTIVATION OF THORACIC GLAND. There was thus clear evidence that both the brain and the prothoracic gland are concerned in the control of growth and moulting in the Lepidoptera. It was therefore suggested by Piepho (1942) that the

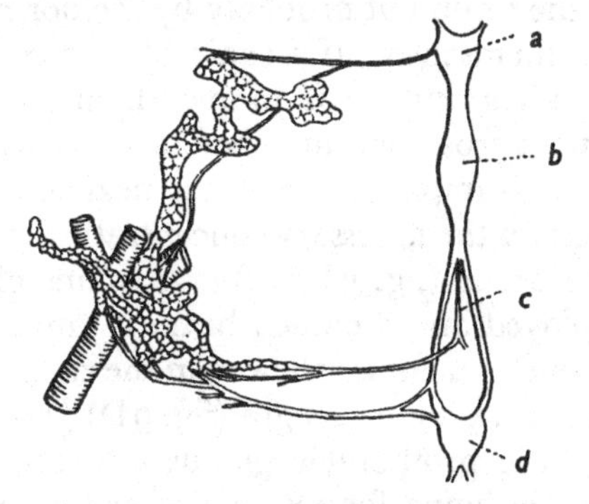

Fig. 10. Prothoracic gland and its nerve supply in *Saturnia*. *a*, suboesophageal ganglion; *b*, prothoracic ganglion; *c*, median nerve; *d*, mesothoracic ganglion. (After Lee, 1948.)

secretion from the prothoracic gland may be produced in response to the secretion from the brain—just as the production of thyroxine by the thyroid gland is secondary to the action of the thyreotropic hormone of the pituitary; and that in the Lepidoptera (as had, indeed, been suggested in *Rhodnius* (Wigglesworth, 1940)), not a single active principle but a succession of active principles is responsible for the moulting process.

It had been noted by Plagge (1938) that although the implanted brain would cause pupation in *Deilephila* larvae from which the brain had been removed, it would *not* induce pupation in the posterior fragments of ligated larvae. The significance of this was proved experimentally by Williams (1947) who repeated the experiment on the diapause pupa of *Platysamia*. A chilled brain implanted in the isolated abdomen of the pupa in diapause will not induce development, whereas it will cause development of the anterior half from which the brain has been removed. On the other hand, the isolated abdomen will develop if it is provided with a chilled brain plus prothoracic glands (fig. 9C). As in *Rhodnius*, the actual source of the secretion in the brain is the neurosecretory cells. There is a medial group of eight cells in each hemisphere and a lateral group of about three cells; they are said to be effective only if implanted together with their organic connexions intact (Williams, 1948*b*).

But the prothoracic glands can be activated not only by the secretion from the brain but probably by the hormone from the thoracic glands themselves. If a chain of isolated abdomens of diapausing *Platysamia* pupae is prepared, in which the body fluid is continuous from one member to the next, and then a prothoracic gland is implanted, only the next adjacent member of the chain receives the necessary concentration of the hormone for moulting to occur (fig. 9E). But if chains of decapitated pupae are employed and a chilled brain is implanted into the first member, development is renewed in the entire series, which may be as much as 24 cm. in length (fig. 9D) (Williams, 1952).

In *Platysamia* the prothoracic glands are still necessary for development in the pupa for six days after the termination of diapause. Thereafter the moulting process can continue independently. In this insect, as we have seen, the arrest of growth, or diapause, in the pupa is due to inactivity in the brain and prothoracic glands. In non-diapausing species there is no such arrest. In most of these both brain and gland remain active after pupation or, if the brain is inactive the gland remains active so that development proceeds in the brainless pupa. This is seen in *Lymantria* (Kopeč, 1922), *Bombyx*, *Galleria* (Bounhiol, 1938), *Vanessa* (Hachlow, 1931), *Phryganidia* (Bodenstein, 1938), *Danaus* and *Prodenia* (Williams, 1952). In some species, on the other hand, the brain remains active but the prothoracic gland is temporarily inactive, and then, if the brain is removed immediately after pupation, no further development occurs. That is so in the bivoltine *Arctias luna* and the polyvoltine *Arctias selene* (Williams, 1952).

It was found by Schmidt and Williams (1953) that spermatogonia and spermatocytes from the testes of dormant pupae of

Platysamia develop promptly into spermatids when placed in a hanging drop of blood taken from insects at a time when the prothoracic gland is secreting its hormone. This provides a very sensitive test by means of which it has been shown that the hormone appears in the larva in a late stage of cocoon spinning and persists for some time after pupation. It is absent during diapause; it then reappears, and it persists in the newly emerged adult in sufficient concentration to cause growth of spermatocytes.

THORACIC GLANDS IN OTHER INSECTS. Glands which are perhaps homologous with the prothoracic glands of Lepidoptera have been found in nearly all the other groups of insects so far examined (Williams, 1949; Pflugfelder, 1949). They are variously placed in the back of the head (ventral glands) or in the thorax (prothoracic, intersegmental, peritracheal, or pericardial glands). The homology of these structures has been proved by embryological investigation only in the Lepidoptera (Toyama, 1902) and Hemiptera (Wells, 1954). In both these groups they arise from the ectoderm of the second maxilla, in close association with the salivary glands. In most Hemiptera they are carried backwards with the salivary glands into the thorax. But, as Wells points out, if this did not happen they would occupy the position of the ventral glands of more primitive groups. We may therefore regard all these glands provisionally as being homologous organs; though it must be pointed out that Pflugfelder (1938*b*) claims that the pericardial gland of *Dixippus* has a quite different origin, from the lateral walls of the coelom.

These glands are always richly supplied by tracheae, and are usually composed of loose or lace-like associations of large cells, often with dendritic or lobulated nuclei (fig. 11) (Toyama, 1902; Pflugfelder, 1938*b*; Wigglesworth, 1952*a*). They always degenerate and disappear soon after or shortly before the completion of the moult to the adult stage—save in the *Thysanura*, which continue to moult in adult life (p. 64) (Gabe, 1953*b*). In Lepidoptera (fig. 10) (Lee, 1948) and in Blattidae (Scharrer, 1948) the glands are richly innervated; the ventral glands of Ephemeroptera, Odonata and the grasshopper *Stenobothrus* receive a nerve supply from the suboesophageal ganglion (Pflugfelder, 1952); but in Hemiptera (Wigglesworth, 1952*a*; Wells, 1954) no nerve supply could be found.

The function of the thoracic gland in those few insects in which it has been studied experimentally appears always to be the same as that described in the Lepidoptera: it is activated

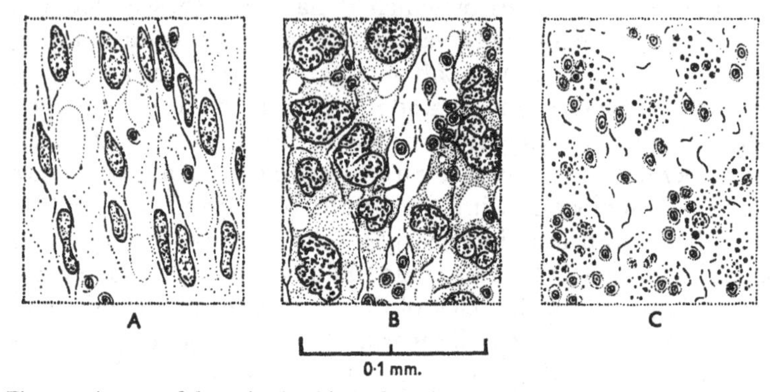

Fig. 11. A, part of thoracic gland in unfed 5th-stage *Rhodnius* larva showing a few haemocytes. B, the same in 5th-stage larva at 10 days after feeding; nuclei of thoracic gland enlarged and lobulated, cytoplasm more dense, haemocytes increased in number. C, the same in adult *Rhodnius* one day after moulting, showing numerous haemocytes around the disintegrating nuclei. (Wigglesworth, 1952*a*.)

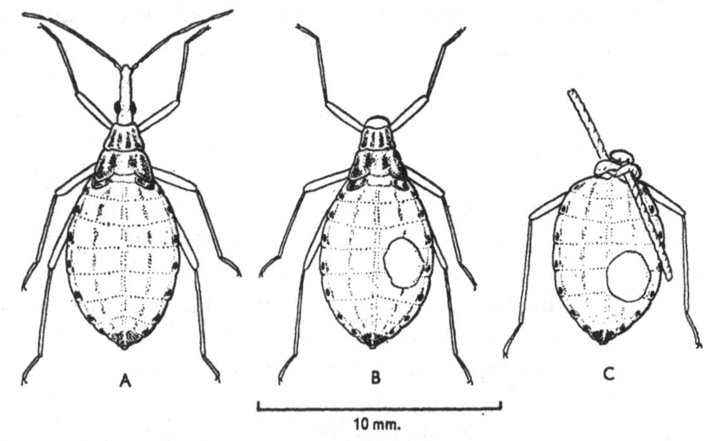

Fig. 12. A, normal 4th-stage larva of *Rhodnius*; B, the same decapitated with implant in abdomen; C, the same ligatured through metathorax with implant in the isolated abdomen.

by the secretion from the neurosecretory cells of the brain and then gives rise to the hormone that induces growth and moulting. Thus, if the brain of a *Rhodnius* larva which has just reached the critical period is implanted into the abdomen of a larva decapitated soon after feeding, it will, as we have seen, cause this larva to moult. But, as in Lepidoptera, it fails to cause moulting in the isolated abdomen (fig. 12). Implantation of the thoracic gland, on the other hand, will induce moulting both in the decapitated insect and in the isolated abdomen (Wigglesworth, 1952*a*).

Similar results have been obtained in *Sialis*. Here the thoracic gland in the full-grown larva appears to have been already activated by the brain in the autumn, but it stores its hormone throughout the winter. The hormone is liberated on warming, so that in this insect pupation will occur even in the decapitated larva when this is brought into warm surroundings (Rahm, 1952). In *Periplaneta* the adult insect can be induced to moult again if larval prothoracic glands (together with corpora cardiaca (p. 64)) are implanted into it (Bodenstein, 1953*b*). In *Locustana* the neurosecretory cells and thoracic glands both go through a cycle of secretory activity during the embryonic moult before hatching from the egg; and if the abdomen is ligatured off the moult is limited to the thorax (Jones, 1953).

GROWTH AND MOULTING IN HIGHER DIPTERA. There is almost no information on the physiological control of the earlier moults in Diptera; but much work has been done on the hardening of the cuticle of the last-stage larva to form the puparium, which immediately precedes the formation of the pupal cuticle within, and on the factors controlling the growth of the imaginal discs—those invaginated parts of the larval epidermis which resume their position in the surface structure at the moment of pupation.

So far as the formation of the puparium is concerned, there is no doubt that the hormonal factor responsible comes from the ring gland of Weismann. If this is excised, the larva of *Calliphora* (Burtt, 1938) or *Drosophila* (Hadorn and Neel, 1938) fails to form the puparium. If ring glands are taken from mature *Drosophila* larvae and used as implants, they will induce puparium formation in larvae before the normal time, or in the ligated hind segments of larvae, or in the male hybrids of the cross *D. melanogaster* ♀ × *D. simulans* ♂, which have a prolonged larval stage (Hadorn and Neel, 1938); and this puparium formation is followed by normal pupation (Vogt, 1942*b*; Bodenstein, 1943*b*).

The ring gland consists of a group of small cells in the mid-dorsal line, which is clearly the corpus allatum; some ventral cells which represent the corpus cardiacum and hypocerebral ganglion; and the large lateral cells which make up the greater part of the organ (fig. 13). It was suggested by E. Thomsen (1942) that these lateral cells are homologous with the 'pericardial gland' described by Pflugfelder in *Dixippus*. That view is now generally accepted. The cells in question have been traced throughout the Diptera, where their association with fairly large tracheae has led to their being named the 'peritracheal gland' (Possompès, 1953). Although the matter has not been proved by embryological studies, all these structures are now usually regarded as homologous with the 'thoracic glands' of other insects as discussed above. Like the thoracic glands they go through cycles of activity in the early larval stages (of *Drosophila*), presumably in association with the moulting cycles (Vogt, 1943*b*), and at metamorphosis they degenerate and disappear.

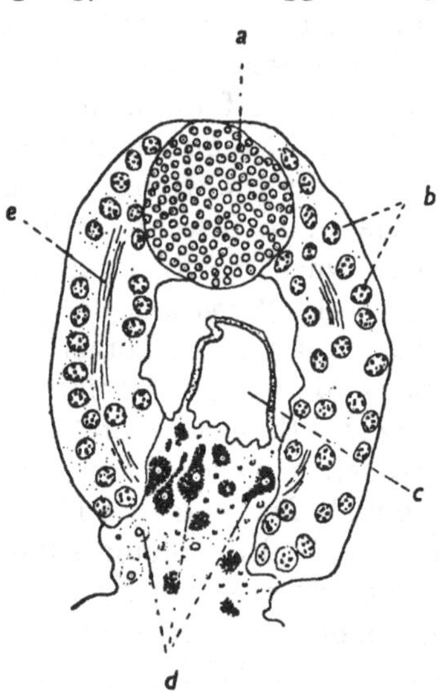

Fig. 13. Section through ring gland of young pupa of *Eristalis* (Diptera) (after Cazal, 1948). *a*, corpus allatum; *b*, large lateral cells ('peritracheal gland'); *c*, aorta; *d*, chromophil and chromophobe cells of corpus cardiacum; *e*, nerve to corpus allatum.

It is these lateral cells of the ring gland which secrete the hormone necessary for puparium formation (Vogt, 1943*b*), and like their presumed homologues, the 'thoracic glands' of other insects, they produce this hormone only after activation by the neurosecretory cells of the brain (cf. de Lerma, 1942). In *Calliphora*, 'persistent larvae' are produced by extirpation of this 'peritracheal gland'; and such larvae have been used by Possompès (1953) to analyse the factors concerned in the secretory activity of the gland. Peritracheal glands transplanted from late 3rd-stage larvae will cause them to pupate, but glands from very young 3rd-stage larvae will not; they must first be activated by the brain.

The method of demonstrating this consists in implanting various structures into the 3rd-instar larva while still in the feeding stage, allowing them to become established, and then extirpating Weismann's ring. If the larva continues its development that must be due to the implant. Isolated ring glands were usually inactive; the brain alone gave negative results; brain and Weismann's ring united by their normal connexions gave 66 per cent positive results, whereas if the connexions were severed only 11 per cent were positive. It appears therefore that activation from the brain takes place by way of the nerve axons (Possompès, 1953).

GROWTH OF IMAGINAL DISCS IN DIPTERA. It will be convenient to consider here the effects of the ring-gland hormone on the growth of the imaginal discs in the Diptera, although we shall return to this question when we come to consider metamorphosis in the next chapter. If the ring gland is excised from the larva of *Calliphora*, or if the gland has had its nerve connexions severed, the larva continues to feed but the growth of the imaginal buds is arrested (Burtt, 1938). The secretion appears to influence growth of the imaginal discs as well as puparium formation. And ring glands of *Drosophila hydei*, taken from larvae ripe for pupation, and implanted into the isolated abdomen of the same species, not only induce formation of the puparium but also cause growth leading to metamorphosis in the gut, ovaries and fat body of the host and in eye-antennal discs or brains transplanted from other larvae (Vogt, 1942 *a*, *b*, 1943 *b*).

The same conclusions have been reached by using the adult fly as the culture medium for imaginal discs. These discs do not continue to grow if transplanted to a normal adult, but they will do so if the larval ring gland is implanted at the same time. By this means it has been possible to show that the gland loses its ability to promote growth in the larval test organs during the first two days of pupal life. This is the time when most of the large lateral cells of the ring gland break down (Bodenstein, 1943 *b*, 1947).

It is clear from these results that the growth of the imaginal discs is dependent on the presence of the growth and moulting hormone. But in the organ discs of the normal *Drosophila* growth appears to be quite independent of moulting; for the curve of growth in the eye discs rises steadily during larval life, with no sign of breaks related with moulting (fig. 14) (Enzmann and Haskins, 1938). (It must be remembered, however, that moulting in *Drosophila* is such a rapid process that cycles of growth in the discs might be difficult to detect.) If there are no interruptions in growth, that would imply that throughout the

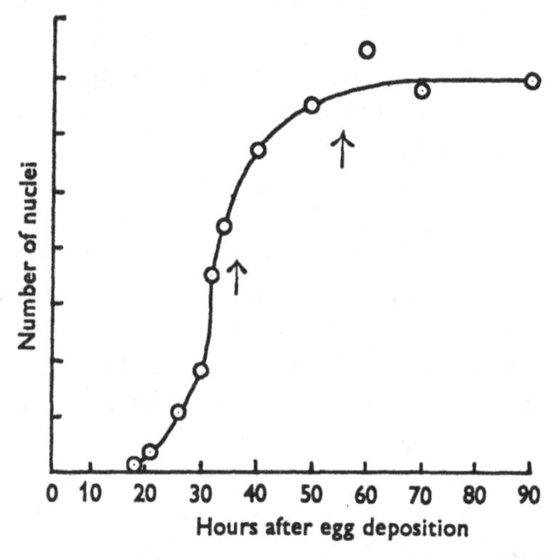

Fig. 14. Growth in cell number in the eye-antennal disc of *Drosophila* larva (after Enzmann and Haskins, 1938). The arrows indicate the presumed times of moulting.

early larval stages in *Drosophila* the growth and moulting hormone is always present in the blood; the imaginal discs can thus grow continuously, but the larval epidermis must go through cycles of cuticle deposition and must therefore of necessity show cyclic growth.

NEUROSECRETION AND ENDOCRINE ORGANS. The first organs to be recognized as ductless glands in the insects were the corpora allata (Nabert, 1913). They arise by budding of ectodermal cells between the mandibular and maxillary segments. Later these cell nests become separated from the epidermis and form compact deeply staining bodies. Sometimes, as in Hemiptera and the higher Diptera, they fuse to form a single median structure. They are innervated from the nerve to the corpus cardiacum. Their function in relation to metamorphosis will be considered in the next chapter.

Closely associated with the corpus allatum is the corpus cardiacum. This organ, bilateral or fused in the mid-line like the corpus allatum, is nervous in origin, arising like the remainder of the stomatogastric nervous system from an ingrowth of the dorsal wall of the stomodaeum. But some of the nerve cells (chromophil cells) may have a neurosecretory function, and in addition there are chromophobe cells, sometimes said to be of neuroglial origin, which also are concerned in secretion.

Little is known of the function of the corpus cardiacum. It is innervated by two nerves from the brain, medial and lateral. The medial nerve takes origin from the neurosecretory cells in the pars intercerebralis; the axons turn downwards from their point of origin and cross over the mid-line. The lateral nerve (absent in some groups) also comes from the protocerebrum but does not cross over (Cazal, 1948; Hanström, 1949).

Increasing interest is being taken in the neurosecretory cells, and in recent years the available information on their distribution and cytology has far outstripped our knowledge of their function. Hanström (1940, 1953; and cf. de Lerma, 1951) has shown that it is possible to trace the evolutionary origin of these cells from the lateral frontal organs of the lower Crustacea. In other Crustacea these structures appear as '*x*-organs'—neurosecretory structures adhering to the brain, the secretion of which is conveyed along their axons to the sinus gland. In the Apterygota (for example, in *Petrobius maritimus*) the corresponding structures take the form of two groups of cells on the surface of the protocerebrum, lying within a separate connective tissue sheath, and connected with the brain only through a small aperture in their capsule. In the Pterygota the cells become imbedded in the brain to form the medial neurosecretory cells.

The massed droplets and granules of secretion which fill the neurosecretory cells give them a luminous bluish appearance when viewed against a black background. This same appearance, particularly evident in dark-ground illumination, can sometimes be seen in the axons running from the cells to the corpus cardiacum (e.g. in Diptera (E. Thomsen, 1954)), and the corpus cardiacum itself may have the same characteristics. Indeed, the secretion from the neurosecretory cells passes along the axons and accumulates in their bulbous swollen extremities within the corpus cardiacum.

This secretory material usually stains deep blue with chrome-haematoxylin after oxidation with permanganate (Gomori, 1941), and by means of this staining method it can be followed along the axons to the corpus cardiacum where it collects between the cells. The corpus cardiacum, in fact, seems to serve as a storage organ for this secretion—presumably releasing it or some modified product of it, into the lumen of the aorta as required. In the cockroach *Leucophaea*, if the nerve to the corpus cardiacum is cut on one side, there is a striking accumulation of this blue-staining material above the cut, and it largely disappears from the corpus cardiacum on that side (fig. 15) (Scharrer, 1952*a*, *b*).

No one could fail to be struck by the resemblance between the corpus cardiacum/allatum system in insects and the pituitary gland system of vertebrates (Wigglesworth, 1934). But this similarity becomes more striking the more closely the systems are compared (Hanström, 1941). In both we have an epithelial ectodermal rudiment (the adenohypophysis or the corpus allatum) which may be homologized hypothetically with a cephalic nephridium. This fuses during growth with a nervous rudiment (the neurohypophysis or the corpus cardiacum) to form a secretory complex. Both neurohypophysis and corpus cardiacum are innervated from neurosecretory cells (in the hypothalamus and in the pars intercerebralis of the protocerebrum respectively), and in both systems it is possible to trace secretory material, staining in the manner described above, from the nerve cells to the endocrine gland.

This general process of transport of secretion along the axons to the corpus cardiacum has been observed in many insects: in Orthoptera (Scharrer, 1952*a*, *b*), Ephemeroptera and Odonata (Arvy and Gabe, 1952, 1953*a*, *b*), Lepidoptera (Arvy, Bounhiol and Gabe, 1953), Coleoptera (Stutinsky, 1952; Arvy and Gabe, 1953*c*) and Diptera (M. Thomsen, 1951; E. Thomsen, 1954). But there are many variants in points of detail. In Ephemeroptera, besides the medial neurosecretory cells of the pars intercerebralis there are neurosecretory cells in the glomeruli and corpora pedunculata the axons from which form the lateral nerves of the corpus cardiacum. Here the secretion is most con-

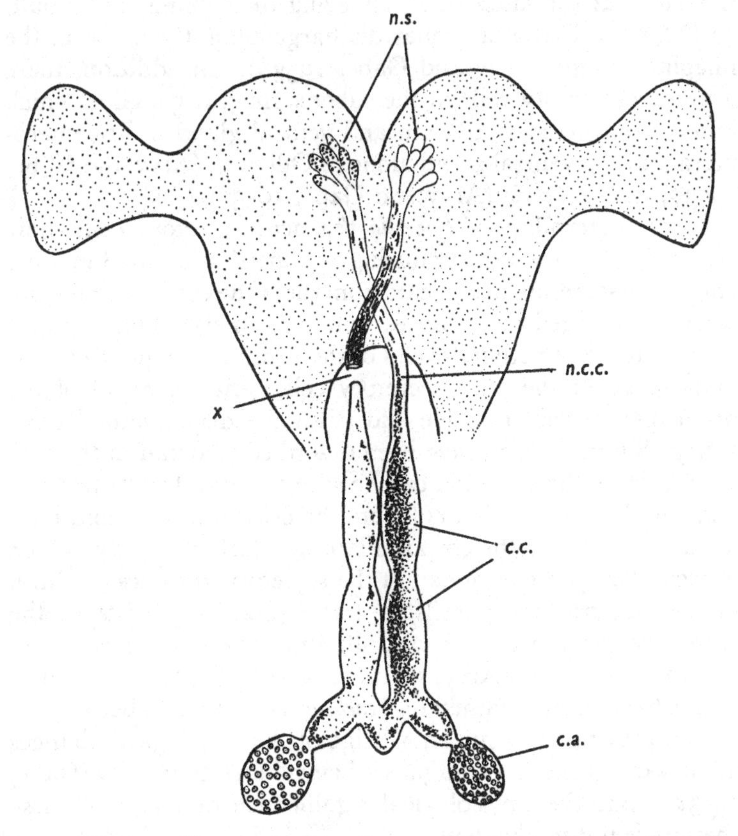

Fig. 15. Diagram of intercerebralis-cardiacum-allatum system in *Leucophaea*. On the left side the nerve to the corpus cardiacum has been cut. Neurosecretory material has accumulated above the point of section and almost disappeared below. On the operated side the corpus cardiacum has slightly decreased in size and the corpus allatum is enlarged (Scharrer, 1952). *c.a.* corpus allatum; *c.c* corpus cardiacum; *n.c.c.* medial nerve to the corpus cardiacum; *n.s.* neurosecretory cells; *x*, point of section of nerve.

spicuous in the neurosecretory cells shortly before the endocrine organs innervated reach their maximum size. At this moment the corpus cardiacum can be seen to be producing acidophil secretion granules which stain red with phloxin and are believed to be discharged into the blood (Arvy and Gabe, 1953*a*). And there seems to be a particularly active period of transport along the axons at the moment of achieving the subimaginal moult. In Odonata there is a similar discharge along the axons at the imaginal moult (Arvy and Gabe, 1953*b*). In addition, there are neurosecretory cells in the suboesophageal ganglion which innervate the corpus allatum and ventral gland in Ephemeroptera and the ventral gland alone in Odonata (Arvy and Gabe, 1953*b*) and *Tenebrio* (Arvy and Gabe, 1953*c*).

In the caterpillar of *Ephestia* the neurosecretory cells show cycles of secretory activity during the larval and pupal moults; and as the secretion disappears from them the corpus cardiacum becomes enlarged. This process comes to an end in older pupae and adults. The activity cycles of the prothoracic glands follow upon those of the neurosecretory cells; after removal of the brain they remain inactive and gradually degenerate (Rehm, 1951). Similar cycles have been described in detail in the silkworm. Here the secretion can be seen to reach the corpus cardiacum about 24–36 hours before the critical period, and later it can be traced to the corpus allatum. During the days which immediately precede pupation the secretory granules diminish in the neurosecretory cells, but the product persists in the corpus cardiacum and corpus allatum. The same cycle is repeated in the pupa (Arvy, Bounhiol and Gabe, 1953). Similar results have been obtained in *Tenebrio* (Arvy and Gabe, 1953*c*). At the present time our knowledge of the histological changes in the various endocrine organs is being added to rapidly (Gabe, 1953*a*); but the physiological significance of many of these changes is not yet known.

We have seen that the product from the neurosecretory cells of the brain serves to activate the thoracic gland and that this activation can usually be brought about even when the brain or the appropriate part of the brain is implanted elsewhere in the body. But we have also seen that in the larva of *Calliphora* conduction by the axons is apparently necessary (Possompès, 1953). It may be that in the normal insect the hormone from the brain always passes along the axons for temporary storage in the corpus cardiacum before being discharged into the blood. That is suggested by the histological evidence, but it has yet to be proved experimentally.

MODE OF ACTION OF THORACIC GLAND HORMONE. The first visible effects of the hormone from the thoracic glands (the moulting hormone) is to cause 'activation' and enlargement of the epidermal cells, followed by mitosis (if growth in size is to occur) and by the secretion of the new cuticle (p. 12). In the special case of puparium formation in the Diptera, the tanning of the larval cuticle is a prominent feature; and this may happen in small patches of the cuticle of caterpillars exposed to very small amounts of the hormone (Kühn and Piepho, 1938).

A highly complex series of metabolic processes has been set in motion. But since respiratory enzymes must play a central part in processes of this kind attention has been focused on the changes which they undergo at this time. Indeed, Dewitz (1905) put forward the view that the oxidases which become active at the time of pupation actually initiate metamorphosis.

The arrest of growth which occurs in diapause may clearly result from the absence or blockage of any of the enzyme systems necessary for growth. It may equally well result from lack of some essential raw material such as salt or water or protein; or from lack of vitamins, some of which play an essential part in those enzyme systems that must become extra active if growth is to occur.

Following upon the observations of Runnström on the changes in the activity of the cytochrome system of the sea-urchin egg at fertilization, Bodine and his co-workers (Bodine, 1934, 1941; Bodine and Boell, 1934) observed that oxygen uptake in the egg of the grasshopper *Melanoplus* is almost completely insensitive to carbon monoxide and to cyanide during diapause, but highly sensitive before and after diapause, whence they concluded that mitosis, growth and differentiation are associated with a functional cytochrome system. They failed to find any close correlation between the intensity of respiration and the amount of cytochrome oxidase (Bodine and Boell, 1936; Allen, 1940); it appeared that some other component of the cytochrome system was the variable factor.

During development in the pupa the intensity of oxygen uptake always follows a U-shaped curve. Throughout pupal development, for example in *Drosophila* (Wolsky, 1938) or *Popillia* (Ludwig, 1953), the respiration is highly sensitive to carbon monoxide and to cyanide (fig. 16). But the course of the curve

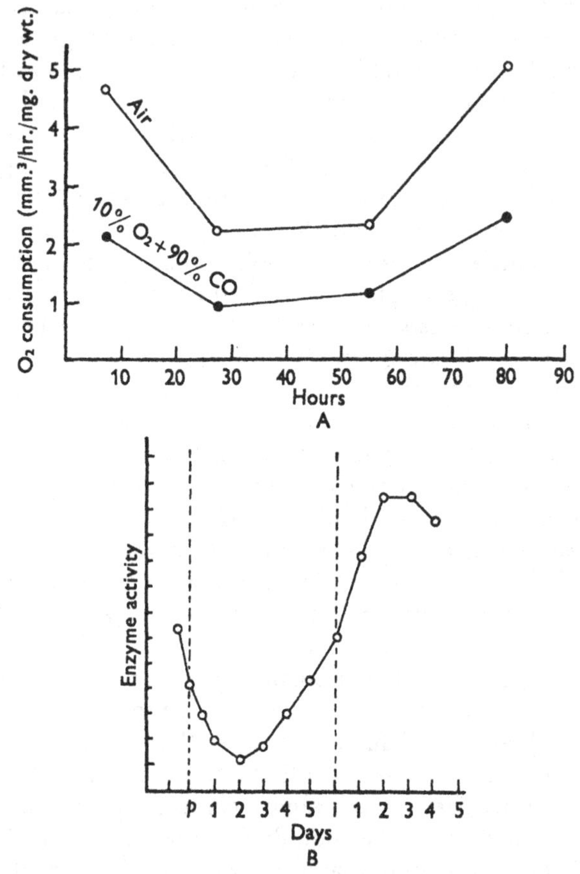

Fig. 16. A, oxygen consumption in pupa of *Drosophila melanogaster* (at 25° C.) in air and in oxygen-carbon monoxide mixture (after Wolsky, 1938). B, activity of cytochrome *c* oxidase in *D. virilis* in pupa and young adult. *p*, time of puparium formation; *i*, time of emergence of adult (after Bodenstein and Sacktor, 1952).

is correlated with the amount or the activity of the Warburg-Keilin system of 'Atmungsferment'+cytochrome (fig. 16B). Either a part of this system is first destroyed (possibly during histolysis) and subsequently rebuilt during the second half of metamorphosis, or else respiratory poisons, which partially inactivate the enzyme system, accumulate gradually at the beginning and later are gradually removed (Wolsky, 1938; Sacktor, 1951; Bodenstein and Sacktor, 1952). The substrate-dehydrogenase systems follow the same U-shaped curve during metamorphosis (Schwan, 1940; Wolsky, 1941), the bulk of the enzyme concerned being probably located in the muscles (Agrell, 1949).

Since the cytochrome system is necessary for active growth and is inhibited by carbon monoxide, exposure of the *Drosophila* pupa or the egg of *Melanoplus* to a gas mixture containing 90 per cent carbon monoxide and 10 per cent oxygen leads to the arrest of development; and since the compound of carbon monoxide with the iron-containing respiratory enzyme is dissociated on exposure to bright light, strong illumination leads to a renewal of growth (Wolsky, 1937; Allen, 1940).

The analysis of the relation of the cytochrome system with growth has been pressed much further by Williams and his colleagues working on the pupa of the giant silkmoth *Platysamia*. The system is highly active in the growing larva which, in consequence, is very sensitive to cyanide and other respiratory poisons. But soon after pupation, when diapause begins, the cytochrome system is virtually out of action. Crystals of potassium cyanide can be implanted without ill effects, except for paralysis of the intersegmental muscles of the abdomen; and spectroscopic examination of these muscles confirms that they alone show the absorption bands of the three cytochromes; the pupal heart continues to beat normally in cyanide Ringer (Williams, 1951). The pupa in diapause shows a similar high resistance to other cytochrome poisons such as diphtheria toxin, which blocks the synthesis of one or more of the components of the cytochrome system. Mature larvae or developing adults have their growth arrested and are killed by 1 μg. of diphtheria toxin; dormant pupae can tolerate 70 μg.; only the intersegmental muscles degenerate (Pappenheimer and Williams, 1952). Imidazole and some of its derivatives such as pilocarpine (all of which combine with haem to form a stable hemichrome and thus render haematin unavailable for cytochrome synthesis) show a similar effect; pupae will live for a year or more after injection of pilocarpine but adult development is prevented (Williams, 1951). Similarly, a prolonged artificial diapause can be induced in fully chilled *Platysamia* pupae by exposure to carbon monoxide (Schneiderman and Williams, 1952).

Throughout diapause in the pupa of *Platysamia* respiration continues at a very low level (Schneiderman and Williams, 1953), and this respiration is resistant to cyanide. The terminal oxidase concerned is probably in part flavoprotein (Chefurka and Williams, 1952), in part cytochrome *e* ($=b_4$) (Shappirio and Williams, 1953). Cytochrome oxidase is present in some quantity, but in the absence of cytochrome *c* it cannot be utilized. One of the earliest signs of the action of the hormone from the thoracic gland, which leads to the termination of diapause and the renewal of growth, is the synthesis of cytochrome *c*. This makes possible the utilization of the cytochrome oxidase and leads to the greatly enhanced respiration associated with growth and mitosis (Williams, 1951; Shappirio and Williams, 1952).

It seems that the complete cytochrome system is not necessary for maintenance but is essential for growth. We saw that the spermatocytes of *Platysamia* can be induced to differentiate into spermatids if cultured in a medium containing the thoracic gland hormone. Such differentiation fails in the presence of such inhibitors of the cytochrome system as carbon monoxide, hydrocyanic acid, azide, diphtheria toxin or pilocarpine, the inhibition by carbon monoxide being reversed by light (Schneiderman, Feder and Ketchell, 1951; Schneiderman, Ketchel and Williams, 1953).

It seems clear, therefore, that the renewal of growth, which in due course leads to moulting, is always associated with synthesis of the complete cytochrome system. Whether or not this synthesis is the prime result of the action of the thoracic gland hormone, or is merely one consequence among many of the renewed growth which this hormone brings about, cannot yet be decided; but it is clearly a change of the first importance. On the other hand, it cannot be claimed that diapause is always associated with a loss of activity in the cytochrome system: this system remains active throughout diapause in the beetle *Popillia* (Ludwig, 1953); and in the sawfly *Pristophora*, which is normally somewhat resistant to cyanide, there is no increase in resistance during diapause (MacDonald and Brown, 1952).

PLATE I

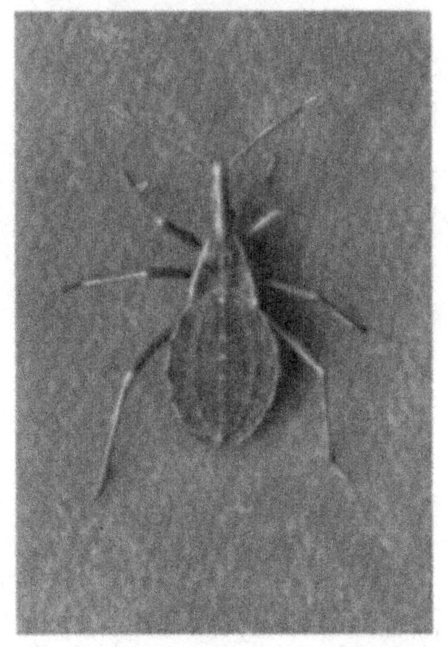

(*a*) *Rhodnius* 4th-stage larva, unfed.

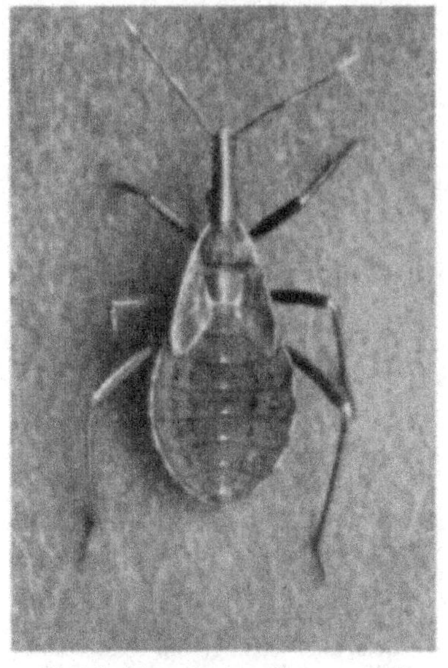

(*b*) *Rhodnius* 5th-stage larva, unfed.

(*c*) *Rhodnius* adult.

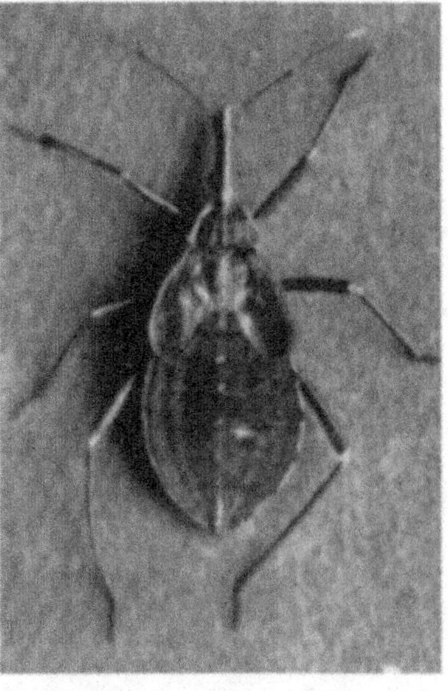

(*d*) *Rhodnius* 6th-stage larva produced by implantation of corpus allatum of a 4th-stage larva into a 5th-stage.

PLATE II

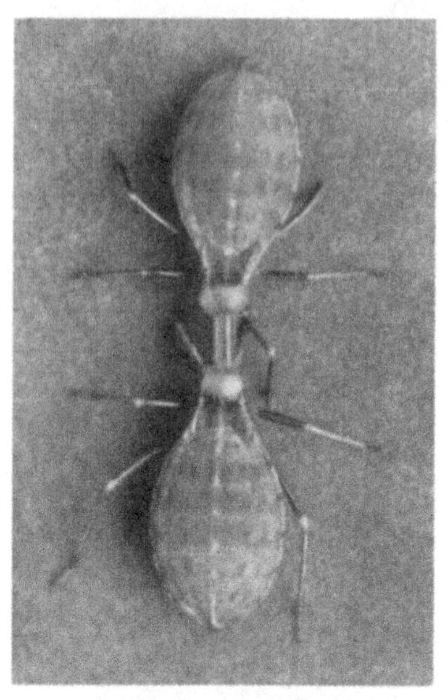

(*a*) *Rhodnius* larva in 4th stage decapitated one week after feeding, and connected by a capillary tube to a 4th-stage larva decapitated one day after feeding.

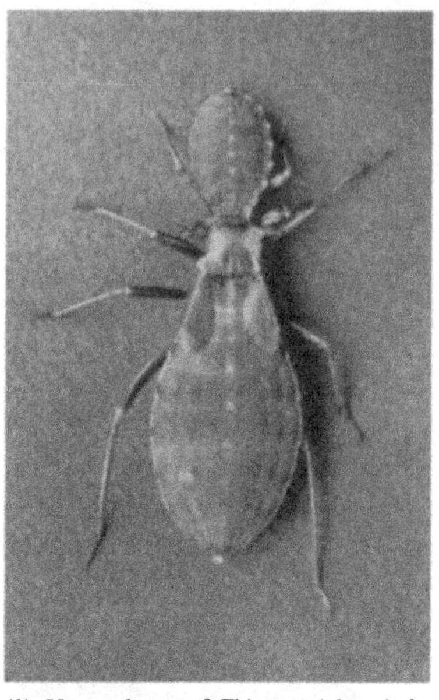

(*b*) Young larva of *Triatoma* (above) decapitated at one day after feeding, in parabiosis with a 5th-stage larva of *Rhodnius*, decapitated ten days after feeding.

(*c*) *Rhodnius* with characters intermediate between larva and adult produced by implanting corpus allatum of a young larva into the abdomen of a 5th-stage larva.

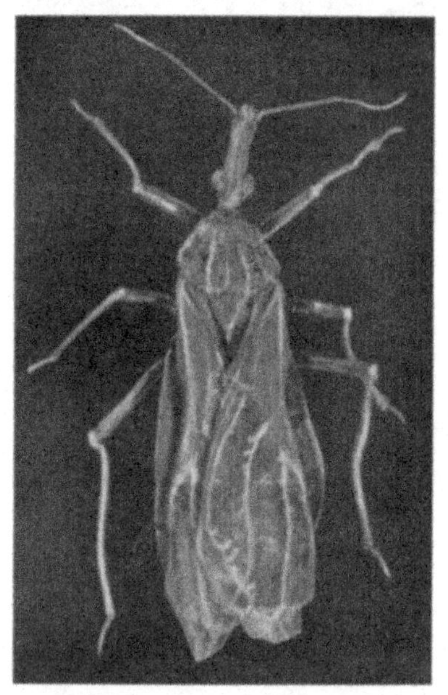

(*d*) Giant adult of *Rhodnius* resulting from the metamorphosis of a 6th-stage larva.

NATURE OF THORACIC GLAND HORMONE. The thoracic gland hormone in insects seems not to be specific: the blood of moulting *Rhodnius* will induce moulting in *Cimex* or *Triatoma* (Plate II *b*) (Wigglesworth, 1936); post-diapause pupae of *Platysamia* will cause renewed growth in diapause pupae of *Telea* to which they are joined (Williams, 1946); extracts from *Calliphora* prepupae will cause puparium formation in *Drosophila* (Karlson and Hanser, 1952); and extracts from pupae of *Galleria* are effective in *Calliphora* (Becker and Plagge, 1939; Becker, 1941). On the other hand, it has been claimed by Vogt (1940, 1941, 1942 *a*) that the ovary of one species of *Drosophila* transplanted into another species will not develop unless the ring gland of the same species is implanted at the same time.

The developing pupa of the silkmoth *Platysamia* has proved excellent material in which to study the properties of the active substance from the prothoracic gland. Using the differentiation of spermatocytes as the method of assay it has been possible to subject the haemolymph to various treatments and then to test again. In this way it has been shown that the active principle is either itself a protein or is closely bound to protein and thus non-dialysable; it is stable at 75° C. for 5 minutes, destroyed at 80° C. for 5 minutes; it remains active for long periods of storage if frozen but is gradually destroyed by repeated freezing and thawing (Schmidt and Williams, 1953). This homogeneous protein fraction has been separated by electrophoresis and shown to combine the property of restoring growth in the diapause pupa with intense agglutination of mammalian red blood corpuscles. Such agglutination can in fact be used as a very sensitive test for the secretion from the prothoracic glands (Ketchel and Williams, 1952).

In recent years highly active extracts of the 'puparium-forming hormone' of *Calliphora* have been prepared. The posterior halves of newly ligatured larvae will readily react to such extracts, but with increasing age they become progressively less responsive; that is, it is the activated epidermis which is responding to the hormone. Whether a single hormone is concerned both in 'activation' of the epidermis and in the subsequent puparium formation, or whether a second hormone is involved, is not yet sure.

At one time it was thought that these active extracts might not contain a true hormone at all, but merely the phenol or phenol precursor concerned in the actual tanning process. But quantitative experiments show that whereas the injection of 100 μg. of pyrocatechol, catechuic acid, or tyrosine are without effect, as little as 0·05 μg. of the purified extract will induce formation of the puparium in *Calliphora*. Conversely, quite high concentrations of the extract do not lead to tanning of the isolated integument. There seems little doubt that we are dealing with a true hormone (Karlson and Hanser, 1953).

Starting with 500 kg. of silkworm pupae, Butenandt and Karlson (1954) have succeeded in isolating 25 mg. of crystalline material, 0·0075 μg. of which was sufficient to induce pupation in a single *Calliphora* larva. When tested by Williams on *Platysamia* this substance proved to be the active principle of the prothoracic gland, for it would induce development in the isolated abdomen of the pupa in diapause. This material is nitrogen free; elementary analysis gives a formula of $C_{18}H_{30}O_4$; and the absorption spectra in the ultra-violet and infra-red suggest that it may be an unsaturated ketone (Karlson 1954).

CONTROL OF HORMONE PRODUCTION. In many insects one cycle of growth and moulting succeeds another without any noticeable pause. In these it is not easy to recognize the nature of the stimuli which bring about the secretion of the hormone concerned. What little information exists on this subject has been obtained in insects with discontinuous growth or insects with well-defined periods of arrested growth or diapause.

In *Rhodnius*, growth and moulting are initiated by the ingestion of a single meal of blood. But a meal that is too small, or a succession of small meals, is not an adequate stimulus, and the larva remains in a state of diapause. It appears to be the stretching of the abdomen and not the state of nutrition which provides the necessary stimulus in this insect. The stimulus seems to be carried to the brain by the nerves; for section of the nerve cord in the prothorax soon after feeding prevents moulting as effectively as decapitation (Wigglesworth, 1934).

There are many examples in other insects of growth being influenced by stimuli which act presumably upon the brain. Most of these concern the control of diapause and are fully dealt with by A. D. Lees (1954) in this series of monographs. For example, many insects utilize the length of day as an index of the approach of autumn. Usually a short day leads to the onset of diapause; a long day, of 16 hours or more, prevents diapause. Presumably the prolonged action of light on the photoreceptors acts by some undetermined route upon the brain and induces this to form or to liberate the activating hormone from the neurosecretory cells (Way and Hopkins, 1950). In the silkworm the 'diapause factor' or 'hibernation substance', which causes the eggs in the ovary to become 'diapause eggs', is derived from the suboesophageal ganglion, but its production or liberation is controlled by the brain acting by way of the oesophageal connectives. The brain exercises the ultimate control (Fukuda, 1951).

Recently, Karlson (1954) has obtained an extract from the heads of adult female silkmoths which acts like pilocarpine and other imidazol derivatives in inhibiting the development of postdiapause pupae (p. 43); but only in those pupae in which the prothoracic gland is still not activated, that is, in which development is still dependent on the secretion of the brain. He suggests that this extract may contain the 'diapause factor' recognized by Fukuda in the suboesophageal ganglion and concludes that this hormone may also be concerned in inducing the pupal diapause of Saturniidae.

Diapause in full-grown larvae of *Lucilia*, induced by desiccation, can be terminated and the larvae caused to pupate simply by placing them in empty glass vials out of contact with sand. Under these conditions the ring gland shows histological signs of renewed activity, but the stimulus has presumably come from the peripheral sense organs via the brain (Mellanby, 1938).

Chelonus annulipes, a braconid parasite of *Pyrausta nubilalis*, never passes beyond the first instar until feeding of its host ceases. It enters a state of diapause which ends only when the host is undergoing changes preliminary to pupation (Bradley and Arbuthnot, 1938). This might be considered as evidence of the growth hormone being effective after absorption by the mouth. But it is equally possible that some other change in the environment is being utilized as a signal by the parasite for the termination of diapause.

Growth in many insects is influenced by what is termed the 'group effect'. There is an optimal density for the development of *Tenebrio* larvae in culture: a moderate degree of mutual stimulation is beneficial; larvae in complete isolation show delayed growth; but above a certain density development is retarded (Hahn, 1932). In *Ptinus tectus* the adverse effect of stimulation is the more evident. If a group of several larvae is reared in one container development may be prolonged by as much as 40 per cent. For example, with a group of eight larvae as much as 5 g. of food per larva is required before the effect disappears; that is, 100 times as much as is needed for full and speedy development of an isolated larva (Gunn and Knight, 1945). The maximum rate of growth in *Blattella* is seen when there is a space of 14 c.c. per insect (Chauvin, 1946). In crowded colonies of caterpillars the time of development may be only four-fifths of that required by solitary individuals, and the number of larval instars may be reduced (Long, 1953). Perhaps these are all examples of the influence of nervous stimulation on hormone production and serve to illustrate the central importance of the brain in the control of growth.

MOULTING OF ADULT INSECTS. We have seen that very soon after the insect becomes adult the thoracic glands break down and disappear. Moulting therefore ceases. But the tissues are still capable of responding to the moulting hormone if they are exposed to it experimentally. The antenna of the adult earwig *Anisolabis* (Furukawa, 1935) or tarsi of the adult *Gryllus* (Sellier, 1946) transplanted to the larva can be induced to moult again. The same occurs when fragments of the integument of adult Lepidoptera are implanted into the abdomen of caterpillars (Piepho, 1938*b*). And the adult bed-bug *Cimex* can be made to moult by joining it to a moulting larva of *Rhodnius* (Wigglesworth, 1940*b*).

The new cuticle laid down by the adult *Cimex* when it moults again is an amorphous structure with no properly formed bristles if an adult several weeks old is used. But if a young adult *Cimex* or an adult *Rhodnius* is caused to moult by joining it to a moulting 5th-stage larva (Wigglesworth, 1940*b*), or by implanting into the abdomen the thoracic glands taken from a 5th-stage larva around the critical period (Wigglesworth, 1952*a*), it lays down a cuticle in which the pattern and structure and the form of the bristles are indistinguishable from those of a normal adult (fig. 17). The adult insect does not have the necessary provision in the structure of the cuticle in the head and thorax to enable it to shed the cuticle when it moults; it remains enclosed within the old skin. But it has even been possible to induce the adult to make two such moults (Wigglesworth, 1940*b*). Whereas

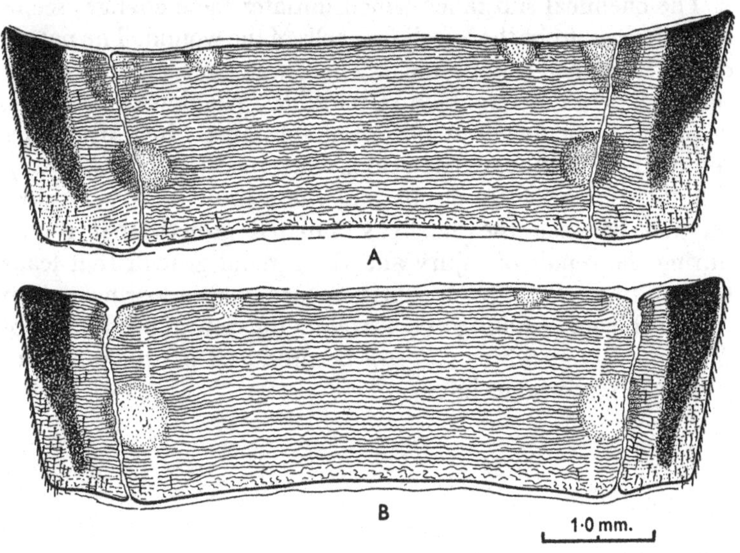

Fig. 17. A, tergite of fifth abdominal segment of adult *Rhodnius*. B, new cuticle of the same segment after moulting induced by parabiosis with 5th-stage larvae. (Wigglesworth, 1940*b*.)

the adult will react in this way to implantation of the thoracic glands it naturally fails to respond to implantation of the pars intercerebralis of the brain (Wigglesworth, 1952*a*).

WOUND HEALING AND MOULTING. If a small area of the integument is excised from the abdomen of an adult *Rhodnius* or from a larva in which normal growth and moulting have been prevented by decapitation, the wound is duly repaired. The cellular changes which accompany such a repair are the same as those which take place in normal moulting (p. 12): 'activation' of the nuclei and enlargement of the cells, mitosis and migration of cells with chromatolysis of the excess nuclei, mutual arrangement of the cells to form a new epidermis, and finally the secretion of new cuticle (Wigglesworth, 1937). And just as normal growth is associated with a complete and active cytochrome system, so also is wound healing; if wounds are made in pupae of *Platysamia* which have had their cytochrome system put out of action by the injection of diphtheria toxin (p. 43), they show no repair (Pappenheimer and Williams, 1952).

The chemical substance which initiates these changes seems to be derived from the autolysing cells of the wound. The nature of this substance is not known. There was some slight evidence that polypeptides might be responsible (Wigglesworth, 1937), or perhaps ribonucleic acid or some substance which stimulates the synthesis of ribonucleic acid is concerned (Lüscher, 1952*a*). But there seems little doubt that there must be a very close relation between the locally stimulated growth that occurs during the repair of injury and the general growth that leads to moulting. Perhaps one may regard the endocrine system responsible for moulting as controlling the central production of the same factors which can still be produced peripherally and locally in response to injury.

CHAPTER 4

PHYSIOLOGY OF METAMORPHOSIS

THE nature of the integument of arthropods, as consisting of a single layer of epidermal cells which lay down a cuticular covering, was first made clear by Schmidt (1845) and Haeckel (1857). Soon after this the nature of the imaginal discs was elucidated by Weismann (1864), with the result that attention was deflected from the part played in metamorphosis by the general epidermis, and most subsequent discussions on the physiology of metamorphosis centred round the activities of the imaginal discs.

In the special case of the higher Diptera and some Hymenoptera, the imaginal discs do play a central part in metamorphosis. In these groups not only do they give rise to specialized adult structures like eyes, wings, legs and antennae, but they produce the whole of the general surface of the thorax and abdomen; only a few of the internal organs persist from the larva to the imago. But such a state of affairs is quite exceptional. By far the greater part of the body surface of most adult insects is formed by the same cells as have laid down the cuticle of the larva.

It follows that the capacity for polymorphism must reside within the cells of the larva, and that the imaginal pattern must exist in imperceptible form in the larval epidermis. Although this is self-evident, and has already been briefly discussed in relation to embryonic development (p. 3), it will be well to describe some specific examples of it.

THE LATENT IMAGINAL PATTERN. If small areas of the integument in the 4th-stage larva of *Rhodnius* are excised and implanted into a different part of the body surface of another 4th-stage larva, they will heal into their new site and perhaps show little difference from the surrounding cuticle. But when the host insect undergoes metamorphosis, these implants develop the adult pattern and structure characteristic of the part of the body from which they have come (Wigglesworth, 1940*a*). Similar results have been obtained in *Ephestia* (Yosii, 1944).

The special characters of the epidermal pattern may be retained by the cells when they multiply and spread during the healing of a wound. If a burn on the abdomen in the *Rhodnius* 4th-stage larva passes through the marginal black spots on the

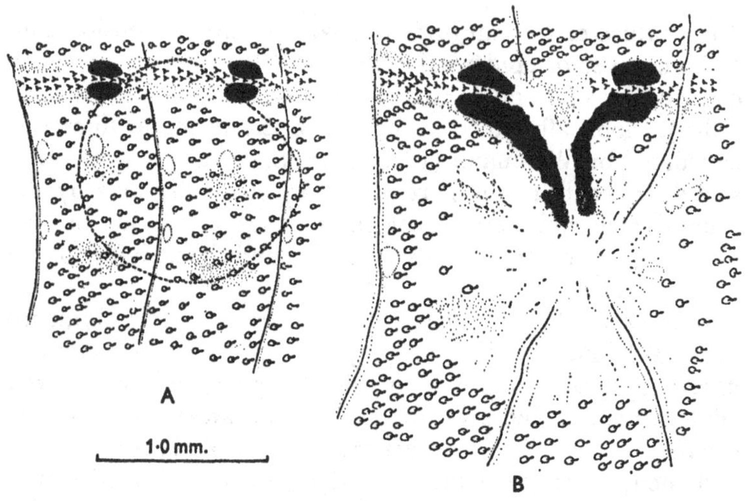

Fig. 18. A, lateral region of the dorsal surface of two abdominal segments in a 3rd-instar larva of *Rhodnius*, showing the double row of marginal plaques above, and beyond this a small piece of the ventral surface. The broken line shows the extent of the epidermis killed by burning. B, the corresponding region in the 4th-instar after healing and moulting. (Wigglesworth, 1940*a*.)

tergites (fig. 18A), the wound is repaired as usual by means of cells spreading inwards from the margin of the burn (Wigglesworth, 1937, 1940*a*). There is, therefore, a migration inwards of cells derived from the region of the black spots. These daughter cells carry with them the potentiality to lay down black cuticle; when the new cuticle of the 5th-stage larva is formed it therefore shows a centripetal displacement of the black pattern (fig. 18B).

If a small burn is applied between the two black spots on the lateral hind-margin of successive segments (fig. 19A, *b*), these are found to have joined up when the insect moults (fig. 19B); if the burn includes the whole of one black spot (fig. 19A *a*), it is eliminated and is absent from the resulting 5th-stage larva (fig. 19B). But the black spots which appear in the adult *Rhodnius* lie at the anterior angles of each tergite, and the posterior angles are pale. Therefore, when the insect just described undergoes metamorphosis at the moulting of the 5th-stage larva, the effects on the pattern are reversed. Where the black spots of the larva had become united the adult black spot has been eliminated, and where the black spot of the larva had been eliminated, the black spots of the adult have become fused (fig. 19C).

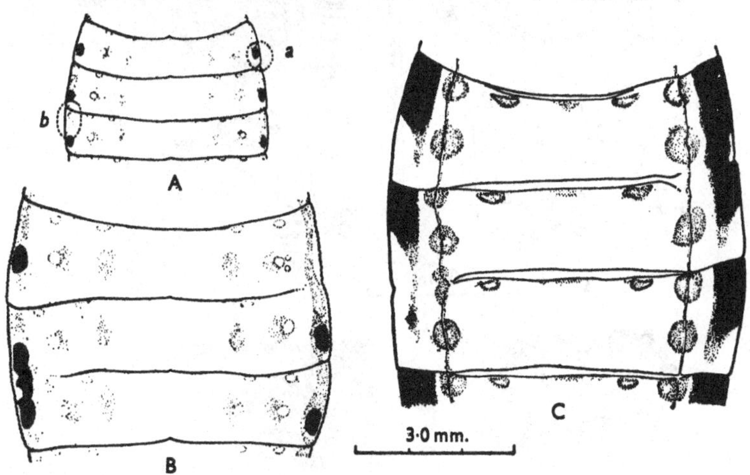

Fig. 19. A, third, fourth and fifth tergites of a normal 3rd-instar *Rhodnius* larva. The broken lines at *a* and *b* show the regions burned. B, corresponding segments in the 5th-instar larva resulting. C, corresponding segments in the adult resulting. (Wigglesworth, 1940*a*.)

The dorsum of the first and second segments in the *Rhodnius* larva does not differ in structure from the rest of the larval abdomen; it has a uniform cuticle with fine stellate folds in the surface, beset with bristle-bearing plaques. But in the adult this region shows the cuticle thrown into characteristic semicircular ridges. If an extensive burn, involving this region of the abdomen, is inflicted on a young larva it may result in little visible effect in the later larval stages (fig. 20A); but when the insect undergoes metamorphosis there is a striking displacement of the ridged pattern (fig. 20B). In all these experiments the migrating and dividing cells, engaged in repairing the injuries, have carried with them the latent potentialities for particular adult structures.

There is no question here of progressive differentiation towards the adult form. There are two forms or patterns, both equally

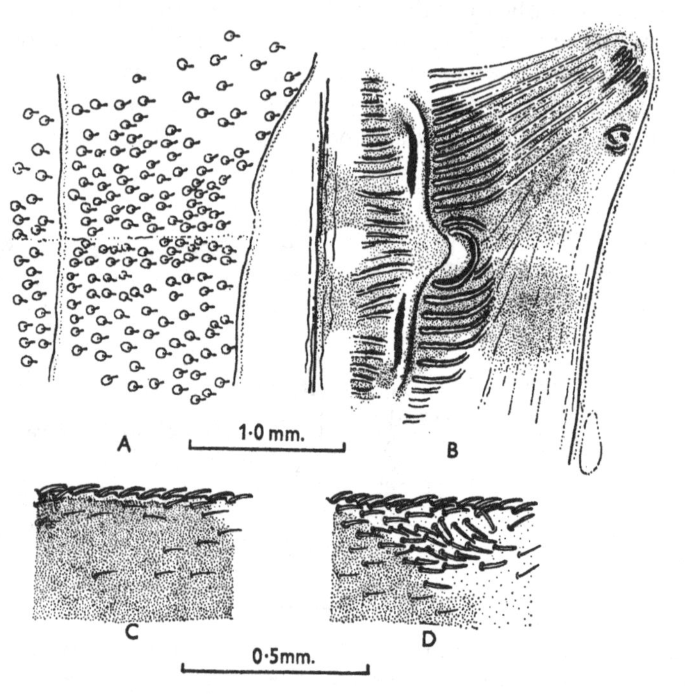

Fig. 20. A, part of mid-dorsal region of first and second segments in a 5th-stage *Rhodnius* larva which had been burned on the right side in the 3rd-instar. B, corresponding segments in the adult resulting. C, marginal spines and part of one of the black spots of a normal adult. D, corresponding region showing marginal spines displacing the black spot in an adult that had been burned in the 3rd instar. (Wigglesworth, 1940*a*.)

specialized; and the one can clearly remain latent in cells which are engaged in producing the other. The problem of metamorphosis is the problem of the control of these alternative patterns.

HUMORAL CONTROL OF METAMORPHOSIS. The first indication of the nature of this control was obtained when it was shown that if larvae of *Rhodnius* in the 1st to the 4th instar are decapitated very shortly after the critical period they often undergo a precocious metamorphosis and develop adult characters in varying degrees (figs. 5, 21) (Wigglesworth, 1934). There is clearly some influence exerted by the head that is preventing the realization of imaginal characters latent within the cells of the larva.

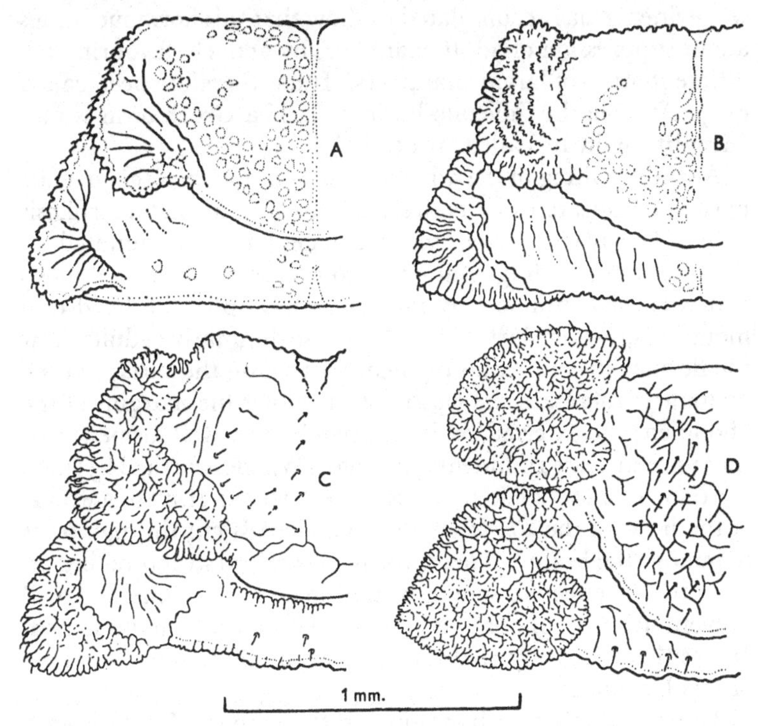

Fig. 21. Dorsal view of the thorax of four *Rhodnius* larvae produced by decapitating 3rd-stage larvae during the critical period (see fig. 5). A, normal 4th-stage larva; B and C, intermediate forms; D, extreme 'adult' form.

This influence was shown to be humoral in nature; for if a 5th-stage larva decapitated at 24 hours after feeding is joined in parabiosis to a 4th-stage larva (7 days after feeding) with the head intact, it is not only caused to moult, under the influence of the growth and moulting hormone from the activated thoracic gland, but its metamorphosis is inhibited and instead of developing adult characters it again lays down a larval cuticle.

The hormone concerned was therefore termed the 'inhibitory hormone'—inhibitory in the sense of preventing the realization of the latent imaginal characters. Unfortunately, the term was understood by some authors in the sense of opposing the action of the growth and moulting hormone. Moreover, as we shall see, evidence has accumulated to show that this hormone brings about suppression of adult characters by actively favouring the differentiation of larval characters. It has therefore been called for preference the 'juvenile hormone'. If a Greek term is preferred it might be called 'neotenin'.

At the moulting of the 5th-stage larva in *Rhodnius* the juvenile hormone appears to be entirely absent so that metamorphosis then takes place. Even the 1st-stage larva, decapitated at 24 hours after feeding, will undergo precocious metamorphosis if it is caused to moult by joining to the tip of the head of a moulting 5th-stage larva (fig. 22). These diminutive adults show the normal structure and pigment pattern on the surface of the abdomen; there are the elastic pleats along the margins of the abdomen; the external genitalia are tolerably well formed; and there are at least rudimentary wings (Wigglesworth, 1934).

CORPUS ALLATUM AS SOURCE OF JUVENILE HORMONE. If the head of the *Rhodnius* 4th-stage larva is cut through so as to remove the brain but leave the post-cerebral structures intact, it is still an effective source of the juvenile hormone and will prevent metamorphosis in a 5th-stage larva to which it is joined. It was therefore suggested that the corpus allatum is the source of this hormone.

This was confirmed by removing the corpus allatum from a 4th-stage larva and implanting it into the abdomen of the 5th-stage larva at 24 hours after feeding. In the most successful experiments these larvae moulted to '6th-stage larvae' (Plate I *d*). In these the leathery wing lobes, though larger than those of the 5th-stage larva, were of the same type; the external genitalia were only slightly more differentiated towards the adult form. In the less successful experiments the 6th-stage insects developed small crumpled wings, and genitalia approaching the adult form, but had larval cuticle all over the abdomen (Plate II *c*). In others the wings and genitalia were of the adult type and the abdominal cuticle intermediate in character. While others were normal adults with just a little patch of larval cuticle over the site of the implant (Plate IV*a*). Such a patch may be quite sharply marked off or it may merge gradually into the imaginal cuticle around (Wigglesworth, 1936).

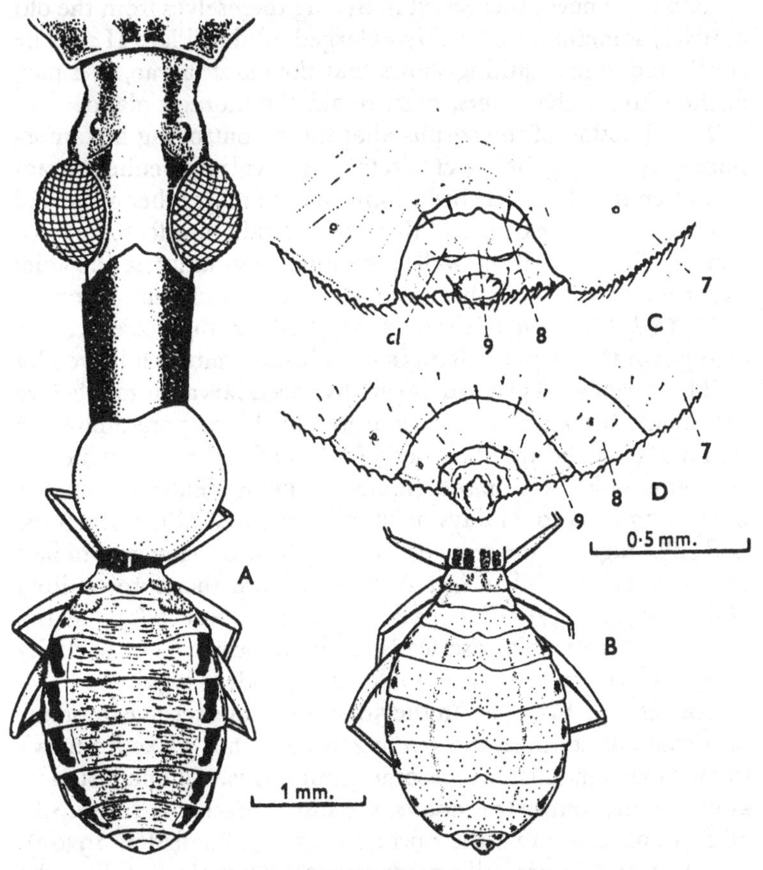

Fig. 22. A, precocious 'adult' *Rhodnius* produced from 1st-stage larva by joining it to the head of a moulting 5th-stage larva. B, normal 2nd-stage larva for comparison. C, terminal segments of precocious 'adult' male. D, the same in normal 2nd-stage larva. 7, 8 and 9 indicate homologous sterna; *cl*, claspers. (Wigglesworth, 1934.)

The same results are obtained if corpora allata from any of the young stages from the 1st to the 4th instar were implanted. But the corpus allatum from the 5th-stage larva had no effect.

Other structures in the head (brain, suboesophageal ganglion, corpus cardiacum, etc.) likewise had no effect.

In some experiments the 6th-stage larvae have fed and moulted again to give rise sometimes to giant 7th-stage larvae (which have never succeeded in freeing themselves from the old cuticle), sometimes to greatly enlarged adults (Plate II*d*). The continuation of moulting shows that the giant larvae, as a part of their larval characters, must retain the thoracic gland.

This function of the corpus allatum in controlling metamorphosis, by causing the insect to retain its larval or juvenile characters when it moults, has been confirmed in most other groups of insects. But in the course of this work certain new features have come to light; it will be well, therefore, to summarize the chief experiments that have been reported in the different orders.

(i) *Orthoptera.* In *Dixippus*, as in *Rhodnius*, there are cyclical changes in the corpora allata in each instar: mitosis followed by visible secretion. There are normally seven larval instars before the female becomes sexually mature. If the corpora allata are removed from the 3rd-stage larva (3 days after the second moult) it moults only twice more (moulting being delayed from the normal 17 days to 25 days or more at the first of these moults, and still longer at the second). It then begins to lay eggs of half the normal size. The last two moults are therefore omitted (Pflugfelder, 1937). Extirpation of the corpora allata in the 1st or 2nd instar likewise leads to the appearance of partially adult characters after two moults (Pflugfelder, 1952).

Conversely, if *Dixippus* in the 5th or 6th stage receive implants of corpora allata from the 3rd or 4th stage they make from two to four extra moults and become giant forms. In one instance, after making four extra moults, the insect reached a record size of 15·1 cm. or double the normal length (Pflugfelder, 1939*a*). Its characters were still predominantly larval; but like the other insects in these experiments it made no further growth although feeding was normal. That suggests that the ventral glands or the pericardial glands had probably either degenerated or become functionless. Indeed, by implanting some ten or twelve pericardial glands into these insects it has been possible to get a few of them to moult with the production of adult characters (Pflugfelder, 1949).

As we have seen, these glands normally degenerate after the last moult; but in the giant *Dixippus* resulting from implantation of corpora allata they persist and hypertrophy enormously, with an increased number of nuclei and the formation of giant nuclei (Pflugfelder, 1939*a*). In fact, as with the thoracic gland in *Rhodnius*, retention of these glands is a juvenile character sustained by the hormone from the corpus allatum. Conversely, when the corpora allata have been removed, and precocious metamorphosis occurs, there is a premature degeneration of the glands (Pflugfelder, 1939*a*).

In the cockroach *Leucophaea* the results resemble those in *Dixippus*. There are usually eight larval stages. If the corpora allata are removed in the 7th stage the insect moults directly to an adultoid form; but if the operation is done in the 5th or 6th stage they make two moults before adult characters are developed (Scharrer, 1946*a*). The fact that in *Leucophaea* (as in *Dixippus*) larval characters are largely retained at the first moult after removal of the corpora allata, and that the thoracic glands remain functional so that a second moult can take place, presumably means that at the time of the operation there is a substantial quantity of juvenile hormone already present in the blood and tissues. The alternative possibility, that the tissues are not yet ready to undergo metamorphosis, will be considered later (p. 75).

In *Gryllus* (Poisson and Sellier, 1947) and in *Melanoplus* (Pfeiffer, 1945*b*) extra larval stages can likewise be obtained by implantation of the corpora allata. *Melanoplus* has six larval stages. The corpus allatum shows histological signs of active secretion all through the earlier larval instars. Secretory activity is evident also in the first part of the 6th stage; but soon the cells shrink and secretion ceases. These observations agree with the experimental results, that in the final larval stage the production of juvenile hormone does not reach a level that is effective in the retention of larval characters (Mendes, 1948).

(ii) *Neuroptera*. In *Sialis*, as in *Rhodnius*, it has been possible to induce precocious metamorphosis by decapitating any of the larval stages for the 1st and the 9th. Extirpation of the corpus allatum alone has the same effect; diminutive adults have been produced from 7th- and 8th-stage larvae (Rahm, 1952).

(iii) *Lepidoptera*. The humoral control of the characters in the integument of the wax moth *Galleria* was demonstrated by the implantation of small fragments into larvae of different age (p. 24). An implant from a young larva, even a newly hatched larva, to a mature larva, pupates prematurely and then forms adult cuticle with scales; while pupal integument implanted into a full-grown larva makes a second pupal moult (Piepho, 1938*a*, *b*).

If the corpora allata are removed from young silkworm larvae in the 4th, 3rd or even the 2nd instar, these proceed to spin cocoons and to pupate at the next moult, and the pupae give rise to diminutive adults (Plate III*c*, *d*, *e*) (Bounhiol, 1938; Fukuda, 1944). The 2nd-instar larvae yield pupae weighing only 25 mg. as compared with a normal pupal weight of 1 g. or more (Bounhiol, 1938). The tiny adults produced are more or less normal in appearance, and they lay eggs, not so large as usual, but not so much reduced in size as the body of the mother. Thus normal moths derived from larvae pupating at the end of the 5th instar laid eggs of an average weight of 0·55 mg., adults from 4th-instar larvae 0·46 mg., and adults from 3rd-instar larvae 0·41 mg. (Fukuda, 1944).

In all larvae after the first instar, pupation can be induced simply by decapitation, if this is done at the proper time. Thus 3rd-instar larvae at 23–26° C. have a normal feeding period of about 80 hours. If decapitated at 35 hours moulting does not occur. If decapitated between 40 and 60 hours they moult to give rise to precocious pupae or to larvae with patches of pupal skin of various sizes (Fukuda, 1944).

Similar results have been obtained with *Galleria*. If the corpora allata are removed from very young larvae they give rise to tiny pupae and imagines; but this does not happen if corpora allata from other larvae have been implanted into the abdomen. Conversely, if corpora allata from young larvae are transplanted to full-grown larvae, larval development is continued with the eventual production of giant pupae and imagines (Piepho, 1943).

The pupal cocoon of *Galleria* is spun by the larva well before moulting begins. Its characteristics are quite different from the web prepared by the larva before a larval moult. The implantation of corpora allata from young larvae into larvae of the final instar, causes them to spin a larval web, an intermediate structure, or a pupal cocoon, depending upon whether they are going to moult later to a larva, an intermediate form, or a pupa. The type of spinning activity is thus determined quite early by the corpus allatum (Piepho, 1950*a*). A similar preliminary or anticipatory action by the endocrine system is seen in the larva of *Cerura vinula* before pupation. These caterpillars cease feeding about 12 days before pupation. Their behaviour changes, they become brown in colour owing to the deposition of ommatine in the epidermal cells, they discharge a red faecal pellet, and they show a marked increase in mitosis in the wing discs. Ligaturing experiments show that these changes are humorally controlled by a factor from the brain which acts upon a thoracic centre. It looks as though the usual moulting hormone system were responsible, acting perhaps at a lower concentration—though it is possible that an entirely separate hormone system may be involved (Bückmann, 1953).

In agreement with the findings in other insects it has been shown that in *Ephestia* there are cycles of secretory activity in the corpus allatum in each of the first four larval stages, but in the last larval stage and in the young pupa the glands appear relatively inactive (Rehm, 1951).

(iv) *Coleoptera.* If corpora allata are taken from young larvae of *Tenebrio* and implanted into larvae in the last instar very soon after moulting, they moult again to produce typical larvae. As many as six extra moults have been obtained, the larvae reaching a very large size (fig. 23A, B) (Radtke, 1942).

(v) *Hymenoptera.* By applying a ligature behind the head in larvae of the honey bee, at the appropriate time, it is possible to get them to moult directly to forms which show certain unmistakably adult characters, such as the formation of a hard exocuticle bearing hairs of adult type and the development of the brush on the first segment of the hind tarsi (Schaller, 1952).

(vi) *Diptera.* Evidence that the corpus allatum is preventing the appearance of adult characters in the young stages of Diptera is not entirely clear-cut. The imaginal discs as they exist in the larva are larval structures; at pupation they undergo metamorphosis, which consists in their differentiation to the adult form.

There can be no doubt that this differentiation is humorally controlled, for it can be brought about prematurely by transplanting the young eye discs of *Drosophila* (from larvae 46–70 hours of age) into larvae that are about to pupate. The implanted discs no longer continue their larval growth but differentiate to form small eyes (Bodenstein, 1939*b*). Conversely, if the eye discs are transplanted into younger larvae their differentiation

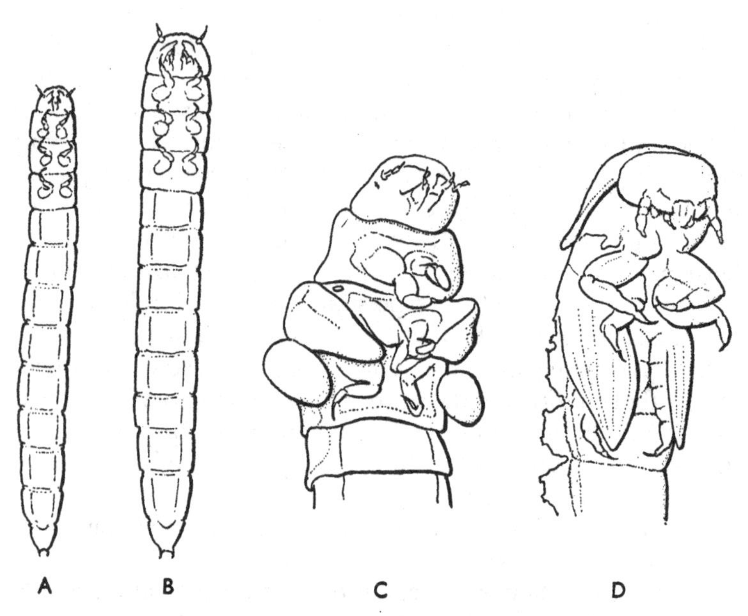

Fig. 23. A, normal full-grown larva of *Tenebrio*. B, giant larva after making four extra moults following implantation of corpus allatum of young larva. C, larva-like intermediate form produced in the same way. D, pupa-like intermediate form similarly produced. (After Radtke, 1942.)

or metamorphosis is delayed; their growth continues longer than normal, and when differentiation finally occurs they give rise to eyes with more facets than usual (Bodenstein, 1939*b*).

It seems probable that this restraint is of the same type as in other insects and is effected by the corpus allatum. This is supported by the fact that the corpus allatum component of the ring gland already shows signs of secretory activity in the 1st larval stage of *Drosophila*, and is much reduced in size, relative to the main part of the ring gland (peritracheal gland), in the last larval stage when the production of juvenile hormone would be expected to fall (Vogt, 1943*a*). Moreover, if the central part of the ring gland (the corpus allatum) is extirpated in the larva of *Drosophila* the development of the eyes is accelerated (Vogt, 1943*b*). We have seen that eye discs implanted into the posterior halves of ligatured larvae of *Drosophila* continue to grow only if ring glands are transplanted at the same time (p. 35). If only the side lobes of the ring gland are implanted the eye discs proceed to differentiate at once; but if the entire ring gland including the central portion (the corpus allatum) is implanted at the same time differentiation is delayed (Vogt, 1943*b*).

We shall see later that in other groups of insects the corpus allatum again produces juvenile hormone in the mature adult. This is so in Diptera also, for if the corpus allatum of the adult *Calliphora* is implanted into the abdomen of the 5th-stage larva of *Rhodnius*, the resulting adult may develop a small patch of larval cuticle over the site of the implant (Wigglesworth, 1954). If the corpus allatum from a *Drosophila* adult is implanted into the larva it exerts a local effect on the cuticle. There is an area of cuticle in the adult which approximates to the larval type (Vogt, 1946*a*). If the same experiment is performed on *Calliphora*, and the corpus allatum from an adult is transplanted to the abdomen of the larva, this shows considerable delay in pupating, and when the puparium is finally produced it is abnormal, being partially larval in structure (Possompès, 1953).

We have already noted several examples of the corpus allatum from one species exerting its effect upon another belonging to a different genus or family. If the decapitated 5th-stage larva of the bed-bug *Cimex* is joined to a 3rd-stage larva of *Rhodnius* with the corpus allatum intact, it develops genitalia which are very incompletely differentiated towards the adult form (Wigglesworth, 1936). The corpus allatum of *Periplaneta* implanted into the abdomen of the last larval stage of the lygaeid *Oncopeltus* (Novák, 1951*a*, *b*) or of *Rhodnius* (Wigglesworth, 1952*b*) becomes duly established and the *Oncopeltus* and *Rhodnius* develop into giant 6th-stage larvae. And the ring gland from the *Calliphora* larva will cause a local formation of larval cuticle in *Rhodnius* (Wigglesworth, 1954). By using the full-grown larva of *Galleria* as a test insect for receiving implants, it has been shown that juvenile hormone is produced by the corpus allatum of *Achroia*, *Ephestia*, *Bombyx mori*, *Tenebrio* and *Carausius* (Piepho, 1950*c*).

DISAPPEARANCE OF THORACIC GLAND. An essential element in metamorphosis is the disappearance of the thoracic gland; for this disappearance ensures that the insect shall not moult again and that the food which would otherwise be utilized for growth or moulting is devoted to reproduction. In the Thysanura, which moult from three to five times a year throughout their adult life, the ventral glands persist (Gabe, 1953*b*); but in all other insects studied those organs which we have grouped together as 'thoracic glands' (p. 31) always degenerate at the time of metamorphosis.

In *Ephestia* this breakdown occurs in the late pupal stage at the time when the gonads are ripening; and it has been suggested that these two changes may be related (Rehm, 1951). In *Drosophila* the side limbs of the ring gland likewise break down in the pupa (Bodenstein, 1947), and in *Calliphora* in the young adult (Thomsen, 1942). In Odonata degeneration may begin within 3 hours after the imaginal moult (Pflugfelder, 1947*b*); in *Forficula* (Lhoste, 1951) and in *Rhodnius* (Wigglesworth, 1952*a*) it is well advanced in 24 hours and complete in 2 days; in *Periplaneta* the prothoracic gland degenerates much more slowly, but breakdown is usually complete by the fourteenth day after metamorphosis (Bodenstein, 1953*b*).

Experiments in *Rhodnius* show that two factors are at work: (*a*) the gland must have experienced the humoral environment of the final moult, that is, in the absence of the juvenile hormone, it must have suffered some change that amounts to 'metamorphosis'; (*b*) it must be subjected to the special conditions, the nature of which is not known, that arise very shortly after the insect moults to the adult. Neither (*a*) nor (*b*) is sufficient by itself to bring about degeneration (Wigglesworth, 1954).

In *Periplaneta*, as in *Rhodnius*, degeneration is prevented by exposure to the juvenile hormone during the period preceding moulting; that is, when an extra larval stage has been produced. But (as in *Rhodnius*) degeneration is not prevented by the implantation of larval corpora allata once metamorphosis has occurred. Neither is degeneration prevented by excision of the corpus allatum and corpus cardiacum from the young adult. On the other hand, if the corpus allatum alone is removed and the corpus cardiacum is left *in situ*, the prothoracic gland remains intact and the adult will moult again (Bodenstein, 1953*b*). The explanation of this interesting result, which, if it proves to be true of insects in general, must hold the key to an important element in the control of metamorphosis, is at present unknown.

CONTROL OF STAGES TOWARDS METAMORPHOSIS. Thus far we have been describing the control of metamorphosis as though this change took place in one single step: larval characters being maintained when growth and moulting proceed in the presence of the juvenile hormone, metamorphosis occurring when the juvenile hormone is absent.

But it was soon realized that that was an over-simplification. Each of the five larval stages of *Rhodnius*, for example, has its own characteristic morphology. The wing lobes grow progressively in each successive instar, the largest change taking place at the moulting of the 4th instar to become the 5th. The sexes become distinguishable by the rudiments of the external genitalia usually in the 3rd instar, always in the 4th; and the differences are conspicuous in the 5th instar (Gillett, 1935). The number of bristle-bearing plaques in the surface of the abdomen increases by about 100 per cent at the first moult, 50 per cent at the second and third moults, 25–30 per cent at the fourth moult, when the 5th-stage larva is produced. (They disappear when the 5th-stage moults to the adult (Wigglesworth, 1940*b*).)

It is evident that, although the metamorphosis that occurs at the fifth moult is by far the most extensive, there is a slight change in the same direction, a very slight differentiation of imaginal characters, at each larval moult; and the extent of this change is greater at the fourth moult than at the earlier ones.

We have seen that in *Rhodnius* (Wigglesworth, 1934), in *Bombyx* (Fukuda, 1944) and in *Sialis* (Rahm, 1952), it is possible to bring about varying degrees of metamorphosis by decapitation soon after the critical period. This was interpreted as meaning that the growth and moulting hormone is liberated before the juvenile hormone; indeed, it is the moulting hormone acting upon the corpus allatum which causes it to secrete its active principle (Wigglesworth, 1936). And since the moulting hormone acting alone leads to the development of adult characters, any delay in the appearance of the juvenile hormone, or any limitation in the amount of this hormone, will permit a small amount of metamorphosis to occur. If there are differences in the concentration of juvenile hormone produced in each instar, or differences in the timing of the secretion, this will result in the morphological changes characteristic of each stage.

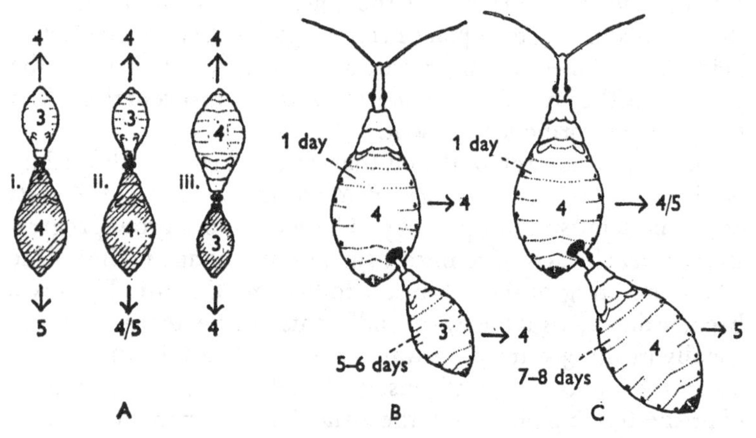

Fig. 24. A, experiments to show the effect of the corpus allatum of the 3rd-stage larva on the characters of the moulting 4th-stage larva. Explanation in the text. The arabic numerals indicate the larval stage. (Wigglesworth, 1936.) B, 4th-stage larva with 3rd-stage larva joined to abdomen. C, 4th-stage larva with a second 4th-stage larva joined to abdomen. The interval since feeding, at the time of joining, is indicated beside each insect in B and C. (Wigglesworth, 1952*b*.)

This conception was tested experimentally by joining different stages of *Rhodnius* in parabiosis. The experiments and the results are summarized in fig. 24A. The shaded insect in each case provides the moulting hormone, i.e. it is decapitated at 6 days after feeding. The black spot indicates that the insect in question, though deprived of its brain, retains the corpus allatum. Females were used throughout. The significant effects were those seen in the 4th-stage larva in experiments (ii) and (iii), in which the insect was exposed to the juvenile hormone secreted by the corpus allatum of the 3rd-stage larva. When it was exposed to the hormone from the outset (iii) it developed again 4th-instar characters; when it received juvenile hormone from the 3rd-stage larva rather later (ii) it developed characters intermediate between a 4th and 5th instar (Wigglesworth, 1936).

These conclusions were based on the minute changes that occur in the rudiments of the external genitalia. But another technique is as follows. A small hole is cut in the abdomen of the insect that is to be subject to experiment; and the head of the insect that is to be used for transfusion, after cutting off the tip, is inserted into this hole and the margin sealed with paraffin wax (fig. 24B). Under these conditions the first insect is often able to free the head and thorax from the old cuticle at moulting and to expand the wing lobes, etc., in a normal manner.

By this means the above results have been confirmed (Plate IV *d*). One must conclude that the corpus allatum of the 3rd-stage larva is adapted to raise and maintain the concentration of juvenile hormone in the blood at a higher level than in the 4th stage. But it was also shown that the *timing* of the secretion is important. For if the 4th-stage larva at 1 day after feeding has joined to it another 4th-stage larva at 7–8 days after feeding (fig. 24C), so that the juvenile hormone is introduced too early in the moulting process, it develops characters intermediate between the 4th instar and the 5th, though approximating more closely to the 4th instar (Plate IV *e*).

The importance of timing in the action of the juvenile hormone is strikingly demonstrated in some remarkable experiments on *Periplaneta*. If two last-stage larvae (presumably of somewhat different ages) are joined together in parabiosis through an opening in the dorsum of the thorax, instead of undergoing a final moult into adults, they make several extra moults, retaining their larval characters to a large degree (Bodenstein, 1953*a*). Whatever the precise mechanism responsible for this upset in the hormone balance, it is evident that the corpus allatum in *Periplaneta* must secrete juvenile hormone at some period during the last instar. It is therefore interesting to note that if an adult *Periplaneta* is caused to moult again, some of the adult structures, notably the 7th sternite of the female, continue to differentiate slightly beyond their normal adult stage; they become 'super-imaginal' (Bodenstein, 1953*a*). We must therefore conclude that *Periplaneta*, like *Gryllus* (Cousin, 1935), normally exhibits a very slight degree of neoteny (p. 70). *Dixippus*, of course, shows very obvious neoteny (Joly, 1945*b*).

CONTROL OF PUPAL CHARACTERS. We thus reach the conclusion that in the hemimetabolous insect the very small steps towards the adult form that appear in the larval stages result from a delicate control of the timing and concentration of juvenile hormone secretion in each instar. That makes it probable that the characters of the pupa in holometabola are controlled in the same way.

We have in fact seen that if the honey-bee larva (Schaller, 1952) is decapitated at the appropriate time, the normal pupal stage may be omitted and incomplete imaginal characters developed directly when the larva moults; that would suggest that pupal characters are differentiated in the presence of some intermediate concentration of juvenile hormone.

By implanting a variable number of corpora allata into the last stage larva of *Galleria*, or by implanting at progressively later days, Piepho (1942, 1951) was able to obtain a graded series of intermediates between larva and pupa; and by implanting corpora allata from full-grown larvae into young pupae he obtained a series of intermediates between pupa and imago. He therefore concluded that before the larval moults there is much juvenile hormone in the body, before the pupal moult less, and before the imaginal moult still less.

This same conclusion, that larval, pupal and adult moults are associated with progressively smaller amounts of juvenile hormone, was reached from a study of the small local effects that may be seen in the cuticle overlying a less successful implant. If a corpus allatum is implanted into the last larval stage of *Galleria* it may give rise to a pupa with larval characters over the site of the implant. When this moults again it may produce an adult with a little patch of pupal cuticle (fig. 25A), an adult with a little patch of larval cuticle (fig. 25B), and in one case there were mixed larval, pupal and adult parts in the same insect (Piepho, 1950*b*). K. K. Nayar (1954), working in my department, has recently shown that the integument of caterpillars, in *Galleria* and other Lepidoptera, if implanted into the body wall of the pupa, will moult directly to produce a cuticle with scales or with structures intermediate between setae and scales, the normal pupal stage having been omitted.

Thus, so far as the control of differentiation is concerned, the pupal stage of a holometabolous insect does not differ from the final larval stage in the hemimetabola. In both cases the characters that appear are controlled by the timing and concentration of the secretion from the corpus allatum.

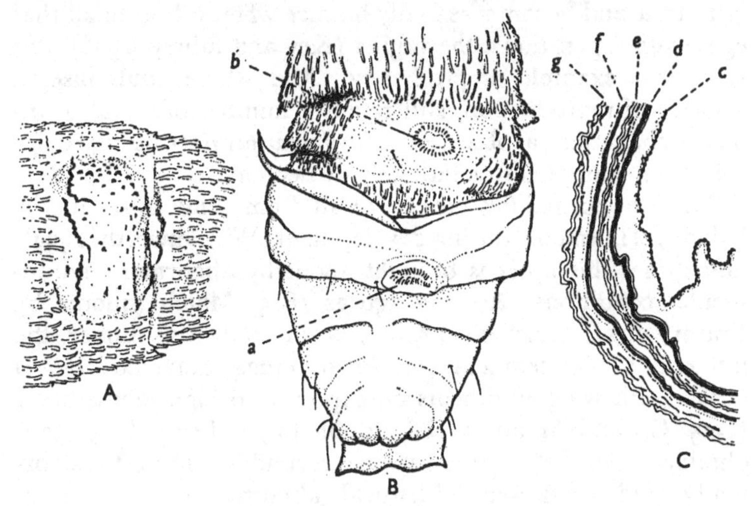

Fig. 25. A, part of abdomen of adult *Galleria* showing a patch of pupal cuticle overlying an implanted corpus allatum. B, part of abdomen of adult *Galleria* with the pupal cuticle still attached. Both the pupa (*a*) and the adult (*b*) shows a patch of larval cuticle with part of a larval proleg over the site of implantation of a corpus allatum (after Piepho, 1950*b*). C, part of an epidermal vesicle formed from larval integument of *Galleria* implanted into last-stage larva and reimplanted into last-stage larva when the first host became adult. *c*, original larval cuticle; *d*, first pupal cuticle; *e*, first imaginal cuticle with scales; *f*, second pupal cuticle; *g*, second imaginal cuticle with scales (after Piepho and Meyer, 1951).

PROTHETELY AND METATHETELY. In some experiments, as we have seen, by upsetting the timing or concentration of the secretion from the corpus allatum, intermediate forms may be produced. It is not an uncommon thing in artificial cultures of certain beetles and Lepidoptera for a few of the larvae to develop visible wing pads or other pupal structures. These monstrous individuals are sometimes regarded as larvae showing premature formation of pupal characters (prothetely), sometimes as pupae whose metamorphosis has been imperfect (metathetely or neoteny).

When the insects in question have an indefinite number of instars and are prone to pupate over a wide range of size, there is little real distinction between prothetely and metathetely. On the other hand, when an insect like *Locusta*, which normally moults five times, develops incompletely adult characters in the 4th instar and becomes sexually mature when still so small that eggs cannot pass down the vagina (Key and Edney, 1936), this is a clear example of prothetely. And where adult insects, capable of reproduction, still retain a number of larval or infantile characters, as observed in *Gryllus campestris* (Cousin, 1935), this is an example of metathetely or neoteny.

These abnormalities clearly result from some upset in the balance of hormones during development (Wigglesworth, 1934). Sometimes this upset is brought about by abnormal environmental conditions. As long ago as 1813, Majoli (quoted by Pruthi, 1924) observed prothetely in silkworms under the influence of high temperature. Pruthi (1924) obtained *Tenebrio* larvae with wing rudiments on exposure to high temperature (29·5° C.) and Arendsen-Hein (quoted by v. Lengerken, 1932) obtained metathetely in the same insect after transfer to abnormally cold conditions. Identical abnormalities have been produced by implanting corpora allata from young larvae of *Tenebrio* into larvae in the last instar (fig. 23) (Radtke, 1942). Nagel (1934) obtained the same results with temperature in *Tribolium*. And intermediate forms between larvae and pupae sometimes appear in *Sialis* if the larvae are kept in the laboratory during the winter (Du Bois and Geigy, 1935).

The effects of temperature on the hormone balance have been studied in *Rhodnius* by placing 4th-stage larvae, immediately after feeding, in a humid atmosphere at temperatures ranging from 18 to 37° C. Fig. 26 shows the number of days after feeding at which moulting took place. At low temperatures the process is greatly retarded. Above 30° C. moulting is again slightly retarded and there is a large amount of individual variation. At 34° C. about 10 per cent of larvae fail to moult; at 35° C. about 50 per cent fail to moult; and at 36° C. and above, none moults. (The lethal temperature is about 40° C.) All the forms produced at moulting are 5th-stage larvae; but if the structure of the wings and genitalia is closely examined it is found that those produced at low temperatures are very slightly more 'juvenile' in character than those of the normal 5th-stage larva; that is, they vary very slightly in the direction of the 4th instar; whereas those produced at the high temperatures are very slightly adultoid. At 35° C. all those which moult show slight adultoid character.

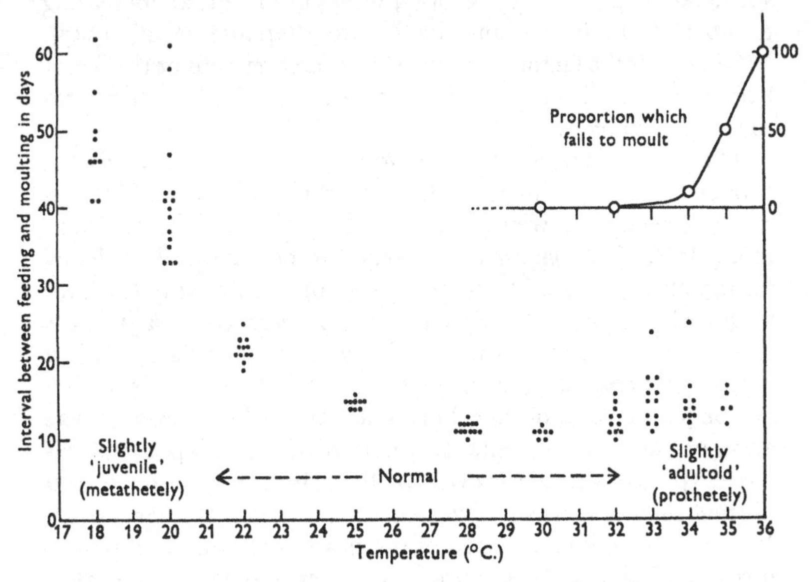

Fig. 26. Effect of temperature on the moulting of 4th-stage larvae of *Rhodnius*. (Wigglesworth, 1952*b*.)

Thus the larvae moulting at low temperatures exhibit slight 'metathetely', those moulting at high temperatures exhibit 'prothetely'. Clearly, low temperature upsets the hormone balance very slightly in favour of the juvenile hormone; while high temperature upsets the balance very slightly in favour of the moulting hormone. But the retardation of moulting as the temperature rises suggests that the secretion of the moulting hormone is itself being depressed. Above 35–36° C. the secretion of this hormone fails entirely, although the other metabolic processes in the insects are not visibly affected (Wigglesworth, 1952*b*).

There are certain other observations which indicate that the secretion of the growth and moulting hormone may fail when the temperature is raised. Thus larvae of *Lucilia* exposed during the prepupal period to a temperature of 35–37° C. are caused to enter diapause; but pupation occurs in about 3 days after return to a lower temperature (Mellanby, 1938). And if post-diapause larvae of the wheat-stem sawfly *Cephus cinctus* in the spring are exposed to 35° C. for 4 weeks or longer, an increasing number of them are put back into diapause (Salt, 1947). Whether it is the factor from the neurosecretory cells or the factor from the thoracic gland which is affected by the high temperature is not known.

The study of prothetely serves to emphasize how precisely the proper hormone balance is maintained in most normal insects. In some species, however, an upset in the balance is very common. In *Gryllus campestris* there seems to be a natural tendency for the juvenile hormone to be slightly predominant in the later stages. Brachypterism, which may be regarded as a juvenile character, seems in this insect always to be associated with a prolonged arrest of development in the 9th stage, that is, with a deficiency of the moulting hormone. If the brain from larvae at an earlier stage is implanted, not only is this diapause eliminated (p. 28) but the growth of the wings is stimulated and normal macropterous individuals are produced (Sellier, 1949). In a few cases, implantation of the brain into a larva of *Gryllus* upset the balance in the other direction and led to a partial precocious metamorphosis (Sellier, 1951).

In Lepidoptera prothetely is particularly liable to occur in species hybrids; for example, in the hybrid *Smerinthus ocellatus* ♂ × *Amorpha populi* ♀; again, presumably, because the hormone balance is upset (Cockayne, 1941). In larvae of *Simulium* infested by the nematode *Mermis* the imaginal discs and the sexual organs are inhibited although the larva continues to grow. It was for this abnormally retarded condition that Strickland (1911) proposed the term 'metathetely'. 'It is evident', he wrote, 'that the cells of the histoblasts are caused to develop by a different stimulus from that which causes development of the larval organs.... This suggests that there may be in the insect body two sets of enzymes which one may term "larval" enzymes and "imaginal" enzymes, and that these may be accelerated or retarded by different factors.' This was an interpretation of prothetely and metathetely which comes rather close to that we have reached to-day.

HORMONE BALANCE IN NORMAL DEVELOPMENT. We have now reached the conception that development in the insect is controlled by the changing balance between the growth and moulting hormone which initiates growth and favours the differentiation of imaginal structures and the juvenile hormone which favours the differentation of larval structures. Metamorphosis results from a relative decrease in the activity of the juvenile hormone.

This conclusion agrees with the histological evidence. In Lepidoptera (*Pieris*, *Lymantria*, etc.) both corpus allatum and prothoracic gland show cyclical changes during moulting, but the relative size of the two organs changes markedly in the course of development. In the newly hatched larva the prothoracic gland is estimated to be twice as large as the corpus allatum; in the full-grown larva it is said to be twenty-nine times as large (Kaiser, 1949) (in view of the highly diffuse character of the prothoracic gland exact comparison can hardly be possible). These observations would support the view that there is a change in the quantitative balance between the two secretions in the course of development. In the pupa there are again signs of secretion in both organs, but these persist far longer in the prothoracic gland (Kaiser, 1949). Likewise in *Drosophila*, the corpus allatum and the side limbs (the peritracheal gland) of the ring gland increase in size in about equal proportions during the greater part of larval life; whereas in the mature larva the peritracheal gland increases much more rapidly (Vogt, 1943*a*).

Bodenstein (1953*d*) even goes so far as to suggest that the corpus allatum secretes the juvenile hormone continuously, but that at metamorphosis the increased activity of the thoracic glands raises the growth and moulting hormone to a concentration sufficient to override the effect of the juvenile hormone and thus brings about metamorphosis in spite of the continued activity of the corpus allatum. Even if that is true in *Periplaneta* with which Bodenstein was concerned, it can hardly be true of insects in general. For implantation of the corpus allatum from the 5th-stage larva of *Rhodnius* into another 5th-stage larva has no effect, whereas implantation of the much smaller corpus allatum from a 4th-stage or a 3rd-stage larva prevents metamorphosis and causes the 5th-stage to develop into a 6th-stage larva (Wigglesworth, 1936).

A progressive change in the activity of the corpus allatum can be seen during the last larval stage in the silkworm. If two or three prothoracic glands are removed from a mature larva and implanted into the abdomen of a young 5th-stage larva within a day after moulting, it either undergoes an extra larval moult or gives rise to an intermediate or metathetelic form (Plate III *b*). If the corpora allata are first excised from such larvae they all pupate normally. Clearly at this period in the instar the corpus allatum is still secreting juvenile hormone in much higher concentration than is needed to ensure the differentiation of the pupal form. Its activity must become reduced in the course of the instar before normal pupation occurs (Fukuda, 1944) (cf. *Periplaneta*, p. 67).

It was observed in *Rhodnius* that if the corpus allatum from a 5th-stage larva or from a very young adult was implanted into a 3rd- or 4th-stage larva, this might show 'prothetely' on reaching the 5th instar and develop wing lobes and genitalia showing some degree of differentiation towards the adult form (Plate IV *b*) (Wigglesworth, 1948 *a*). Now it was shown by Pflugfelder (1939 *a*) in *Dixippus* and by Bodenstein (1947) in *Drosophila* that the implantation of extra corpora allata leads to the partial atrophy of the corpus allatum of the host. And it has been observed in *Rhodnius* that after one or two moults a corpus allatum implanted into the abdomen also shows partial atrophy with chromatolysis of the nuclei (Wigglesworth, unpublished). It may be that if the insect's own corpus allatum has partly degenerated meanwhile, there will be a slight deficiency of juvenile hormone with consequent prothetely. But as an alternative hypothesis, which is perhaps worthy of further test, it was suggested that in the final instar the corpus allatum not only ceases to secrete the juvenile hormone but actively removes from the blood any of the hormone already present in it (Wigglesworth, 1952 *b*).

In *Dixippus* it has been possible to upset the hormone balance

PLATE III

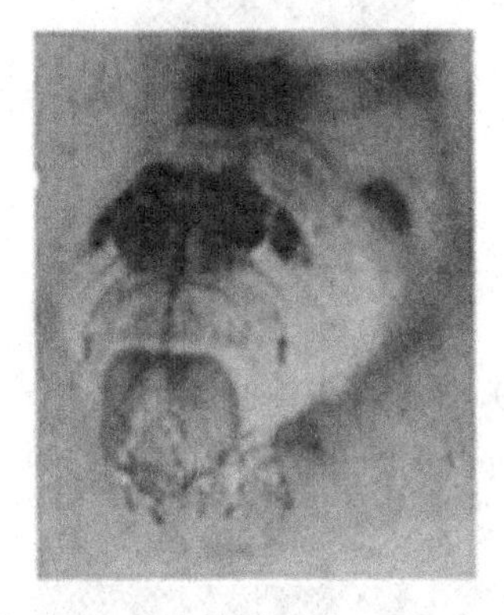

(*a*) Front view of normal mature larva of silkworm (5th stage). (After Fukuda.)

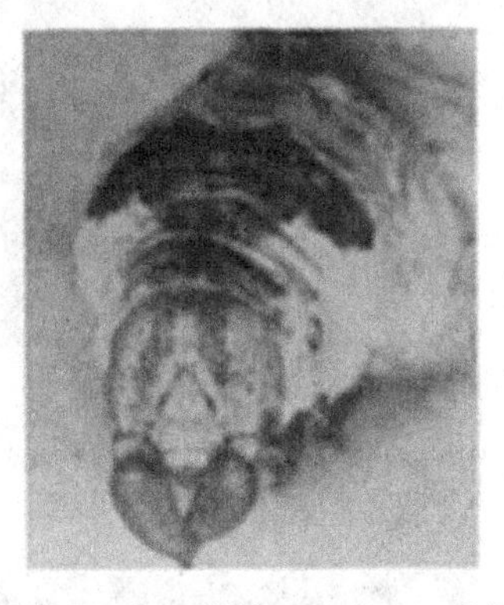

(*b*) Front view of 6th-stage larva of silkworm produced by the implantation of prothoracic glands into a 5th-stage larva early in the instar. It has developed pupal antennae. (After Fukuda.)

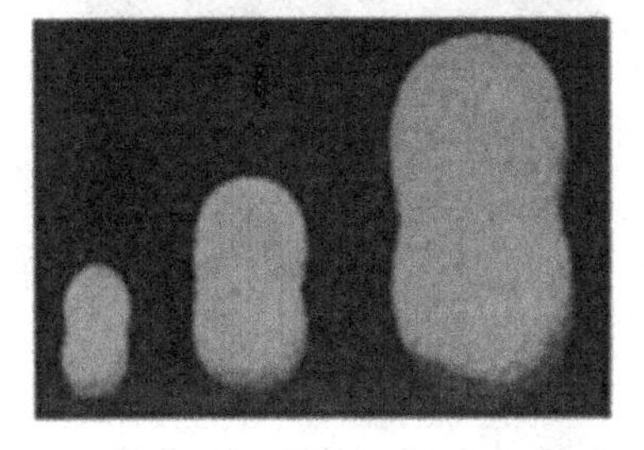

(*c*) Cocoons of the silkworm: from larva in which the corpora allata were extirpated in the 3rd stage; from larva similarly treated in the 4th stage; from normal larva pupating after the 5th stage. (After Fukuda.)

(*d*) Pupae from the cocoons shown in (*c*). (After Fukuda.)

(*e*) Silkmoths derived from the pupae shown in (*d*). (After Fukuda.)

PLATE IV

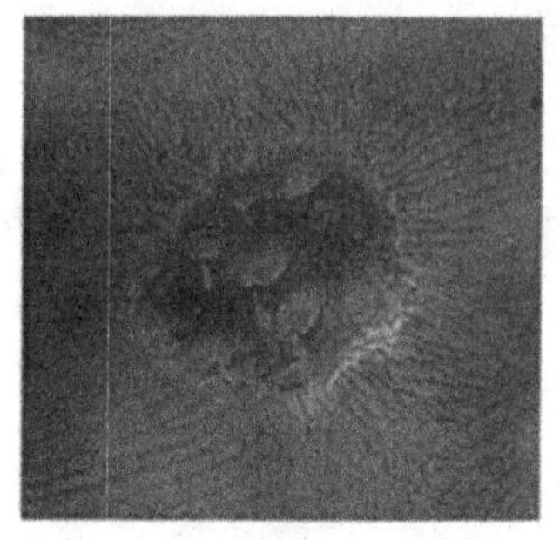

(*a*) Small patch of larval cuticle on the abdomen of an adult *Rhodnius*, overlying an implanted corpus allatum.

(*b*) *Rhodnius* 5th-stage larva showing 'adultoid' characters, following the implantation of the corpus allatum of a newly moulted adult into a 4th-stage larva.

(*c*) Normal 4th-stage larva of *Rhodnius*.

(*d*) *Rhodnius* 5th-stage larva produced from a 4th-stage to which a 3rd-stage had been joined (fig. 24B). It has developed 4th-instar characters again.

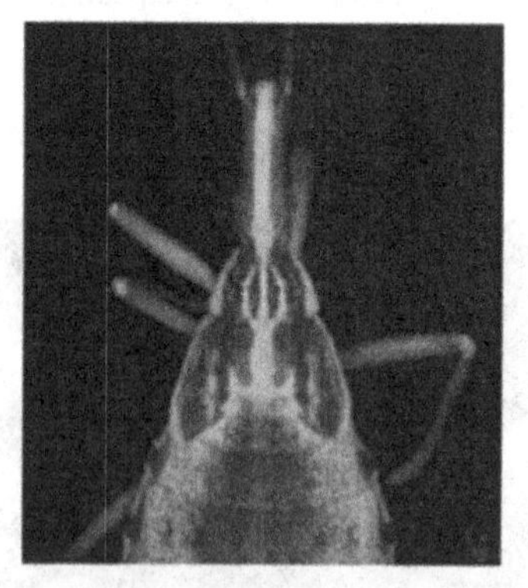

(*e*) *Rhodnius* 5th-stage larva produced from a 4th-stage to which another 4th-stage larva seven days after feeding had been joined (fig. 24C). It has developed characters intermediate between a 4th- and a 5th-stage larva.

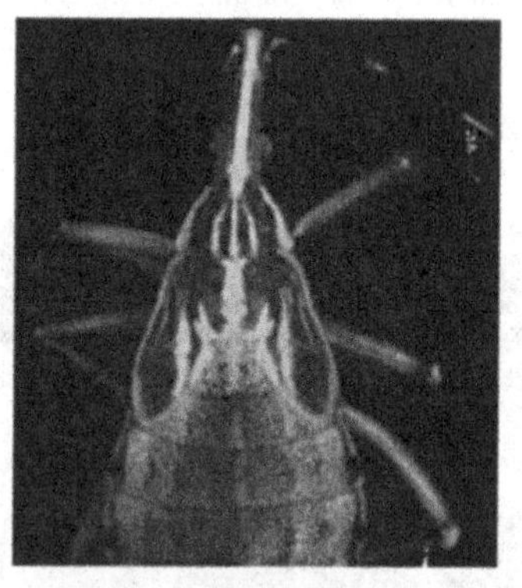

(*f*) Normal 5th-stage larva of *Rhodnius*.

in another way, by extirpation of the pericardial gland. If this gland is looked upon as a part of the thoracic gland system (p. 31) which secretes the growth and moulting hormone, its extirpation will be equivalent to a partial elimination of this system. If this operation is done on the last-stage larva it fails to undergo a proper metamorphosis: the sexual organs of the full-grown insect are more or less larval in character. It appears that the balance has been upset in favour of the juvenile hormone. These results suggest that in this insect (which, as Joly (1945*b*) points out, has a more or less neotenic adult) the corpus allatum is still engaged in secreting juvenile hormone in the last larval instar (Pflugfelder, 1949) (cf. *Periplaneta*, p. 67).

COMPETENCE OF TISSUES TO UNDERGO METAMORPHOSIS. Up to this point we have assumed that provided the hormone balance is favourable the insect may undergo metamorphosis at any stage in its post-embryonic growth. That certainly appears to be true for most of the organs in *Rhodnius*, which can be made to undergo metamorphosis in the 1st instar (p. 56) or in Lepidoptera which, after removal of the corpora allata, will pupate in the 2nd instar (p. 60), and in which fragments of the integument will undergo metamorphosis immediately after hatching from the egg if they are implanted into the last-stage larva (p. 60).

Both *Dixippus* and *Leucophaea* (p. 59), from which the corpora allata have been removed in an early larval stage, undergo two moults before developing adult characters; and this was attributed by Scharrer (1946*a*, 1948) to the tissues being as yet incapable of metamorphosis. But in view of the results on Hemiptera and Lepidoptera it seems more probable that the delay in metamorphosis in *Dixippus* and *Leucophaea* results from the blood already containing some juvenile hormone when the corpora allata were removed (p. 59).

On the other hand, when a multicellular organ reaches a certain size and complexity it will require a certain minimum number of cells in order to achieve normal differentiation. That is apparent in the wings of *Rhodnius*. The general cuticle of the abdomen, and even the genitalia, show good differentiation of adult characters when metamorphosis is induced in the 1st-stage larva; but the wings are just little crumpled lobes. Even in the 4th-stage larva undergoing metamorphosis the wings are a very poor copy of those of the adult. The rudiments of the wings are clearly not fully competent to undergo metamorphosis until the 5th instar. The same deficiency is seen in the differentiation of the ovaries in 3rd-stage larvae of *Rhodnius* when they

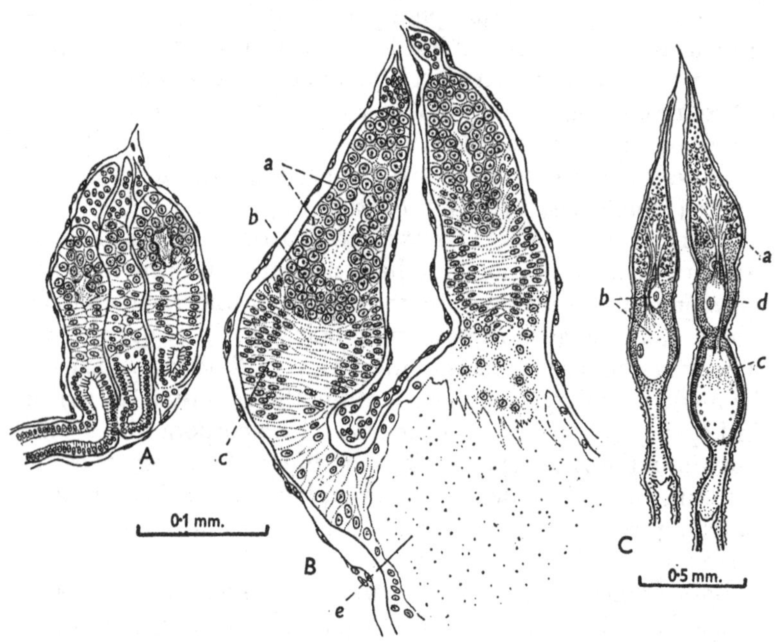

Fig. 27. A, section of ovary in newly moulted 4th-stage larva of *Rhodnius*. B, section of ovary in 'precocious adult' produced from 3rd-stage larva by joining it to a moulting 5th-stage larva. C, section of ovary of normal adult at the time of moulting, showing much more advanced differentiation. *a*, nurse cells; *b*, oocytes; *c*, follicular cells; *d*, nutritive cords; *e*, oviduct. (Wigglesworth, 1936.)

undergo a precocious metamorphosis (fig. 27) (Wigglesworth, 1936). Likewise, the genital rudiments of young *Periplaneta* implanted into the integument of the last-stage larva undergo a complete metamorphosis when the host moults, provided they have been taken from the 9th-stage larva (there are normally ten larval stages). But if they have been removed from a 4th- or 5th-stage larva the development of adult characters is incomplete. The rudiments are not yet fully competent (Bodenstein, 1953*a*).

This relation is even more evident in holometabolous insects in which so large a part of the adult organism is developed from the imaginal discs. In Lepidoptera the immature larval ovary can be made to develop prematurely by transplanting it into a young pupa (Fukuda, 1939). But it must first reach a certain stage of growth (Kopeč, 1911).

The problem has been studied in greatest detail in *Drosophila*. Here also the larval ovaries must reach a certain size before they can undergo the change of metamorphosis (Vogt, 1940). The germs of other organs vary in their capacity to differentiate in response to the hormones of the mature larva: the leg discs respond more easily than the genital discs (Bodenstein, 1943*b*). At a given stage of development the various anlagen, such as salivary glands, eye, genital discs, even various regions within the same rudiment, may differ in their response (Bodenstein, 1943*b*). The salivary gland affords a particularly striking example. If this is transplanted at too young a stage into larvae shortly before pupation, not only does the imaginal rudiment fail to differentiate, but much of the larval tissue still persists when the host becomes adult (Bodenstein, 1943*a*).

There is no doubt that in *Drosophila* and other Diptera, the stage of growth in the imaginal discs is an important factor in imaginal differentiation. But even here the controlling influence of the hormones is evident. The imaginal discs of the last-stage larva of *Aedes* will differentiate in the young pupa, less well in the older pupa, but not at all in the adult (Bodenstein, 1946). If the eye disc is removed from the 3rd-instar larva of *Drosophila* and implanted in the adult fly (in which the juvenile hormone is again active (p. 63)) it will grow considerably before differentiation begins (Bodenstein, 1943*b*); whereas if it is implanted into the mature larva it will cease to grow and begins to differentiate immediately to a small imaginal eye (Bodenstein, 1939*b*, 1941*a*).

ENDOCRINE GLANDS AND REPRODUCTION. Before examining more closely the mode of action of the hormones in controlling metamorphosis we must consider certain other functions exerted by these hormones, and particularly by the secretion from the corpus allatum. Very soon after the adult *Rhodnius* has been fed the corpus allatum shows signs of renewed activity; and if it is removed by decapitation the oocytes of the female fail to develop and the accessory glands in the male do not become filled with secretion (Wigglesworth, 1936).

In the fasting insect, or in the absence of the corpus allatum, the oocytes continue to grow until the point at which yolk should be deposited. Then the nucleus degenerates and the follicular cells lose their regular epithelial arrangement and begin to invade the oocyte. As they move inward they absorb the cytoplasm until finally the interior of the oocyte is completely filled with follicular cells, many of which are breaking down by chromatolysis. Gradually the entire mass decreases in size and ultimately disappears; and the next oocyte to reach the follicular stage suffers the same fate. If the decapitated insect is joined in parabiosis with either a male or another female with the corpus allatum intact, or if a corpus allatum is implanted into the abdomen, the deposition of yolk takes place normally (Wigglesworth, 1936).

Precisely similar results have been obtained in *Melanoplus*, where both yolk deposition and secretion in the oviduct are dependent on the corpus allatum (Pfeiffer, 1939), and in *Leucophaea*, where yolk formation and secretion in the accessory gland can be restored by reimplantation of the corpus allatum (Scharrer, 1946*b*). In *Calliphora*, also, removal of the corpus allatum in the adult fly usually prevents the formation of yolk in the oocytes (Thomsen, 1942); and likewise in *Drosophila* (Vogt, 1943*c*). In *Dytiscus* the extirpation of the corpora allata gradually brings to an end the development of the eggs in the active adult female; and the implantation of corpora allata (of either male or female) induces ovarial activity in the inactive female in diapause. Here again the effect is on the deposition of yolk. It is noteworthy that if the corpus allatum is implanted actually within the ovary of *Dytiscus* it is the oocytes immediately adjacent to it which are most affected (Joly, 1945*a*).

If the ovaries are removed from certain insects the corpus allatum hypertrophies. That is seen in *Calliphora* (Thomsen, 1942) and *Lucilia* (though not in *Sarcophaga*) (Day, 1943). In *Rhodnius* the corpus allatum of a normal adult is about 70 μ in diameter; when the ovaries are removed it is increased to about 100 μ (Wigglesworth, 1948*a*). A similar hypertrophy is seen in the *Drosophila* mutant 'female sterile', and in female hybrids of the cross *melanogaster* × *simulans*, in both of which the ovarioles fail to develop. Whether this hypertrophy signifies some reciprocal action of the ovaries upon the corpus allatum (Thomsen, 1942), or whether it is merely the result of the secretion being retained by the corpus allatum, since it is not required by the ovaries, is not known.

If the corpus allatum of the adult *Dixippus* is implanted into the last-stage larva it causes this to make a series of extra larval moults (Pflugfelder, 1940). Pfeiffer (1945 *b*) obtained the same result in *Melanoplus*. In *Rhodnius* the corpus allatum of the adult will prevent metamorphosis in the 5th-instar larva, and the corpus allatum of the 4th-instar larva will induce egg development in the decapitated adult (Wigglesworth, 1948 *a*). And in *Drosophila* the corpus allatum of the larva, even in the 1st instar, will restore egg development in an adult with the gland extirpated (Vogt, 1943 *c*; Bodenstein, 1947). Moreover, the corpus allatum of the adult *Periplaneta* will prevent metamorphosis in *Oncopeltus* (Novák, 1951 *a*, *b*) and in *Rhodnius* (Wigglesworth, 1952 *b*); and the corpus allatum of the adult *Calliphora*, implanted into the 5th-stage larva of *Rhodnius*, may lead to the formation of a localized patch of larval cuticle when the insect moults to become adult (Wigglesworth, 1954).

It would appear from these results, as was suggested by Pfeiffer (1945 *b*), that the hormone necessary for yolk deposition in the adult is the same as the juvenile hormone. But the situation does not seem to be quite so simple as that. The corpus allatum shows signs of renewed secretory activity in adult Lepidoptera (Ito, 1918); but we have seen that silkworm caterpillars from which the corpora allata have been extirpated immediately undergo metamorphosis to produce moths and that these develop eggs as usual (Bounhiol, 1938; Fukuda, 1944). A certain number of *Calliphora* females deposit yolk in the eggs in spite of removal of the corpus allatum (Thomsen, 1952); and if the gland is extirpated in the last larval stage the resulting adults generally lay viable eggs (Possompès, 1953).

Indeed, it has been found that in *Calliphora*, not only the corpus allatum, but the corpus cardiacum and the neurosecretory cells in the brain can all influence egg development. The neurosecretory cells can act by way of a secretion discharged into the blood; for their removal prevents egg development much more effectively than removal of the corpus allatum, and this function can be made good by implantation of the neurosecretory cells from another fly. The corpus cardiacum has the same effect in less degree; for flies with neurosecretory cells and corpus allatum intact can develop their eggs without the corpus cardiacum; but an implanted corpus cardiacum, provided it has been taken from a mature fly, can replace the neurosecretory cells (Thomsen, 1952). It may be recalled that in the mature fly the secretory product from the neurosecretory cells flows along the axons to the corpus cardiacum (Scharrer, 1953; Thomsen, 1954).

Thus it may be that in *Dixippus*, Lepidoptera and other insects which develop their eggs normally after extirpation of the corpus allatum, the neurosecretory cells in the brain are fulfilling the necessary function. In the adult male *Bombyx* there are few secretory granules in the neurosecretory cells, but in the female the secretion again appears and is transported along the axons shortly before oviposition (Arvy, Bounhiol and Gabe, 1953). A similar intense secretion is seen in the egg-laying female of phasmids (Dupont-Raabe, 1951, 1952), and the neurosecretory cells in the pyrrhocorid *Iphita limbata* show a reduction in the secretory granules after oviposition (Nayar, 1953). The corpus allatum shows histological signs of renewed activity in the adult, both in Lepidoptera (Kaiser, 1949) and in phasmids (Pflugfelder, 1937), secretion being most evident at the time of egg development.

ENDOCRINE GLANDS AND METABOLISM. Every endocrine gland affecting growth exerts its action by influencing metabolism. It may have a general effect on metabolism, or it may affect specifically some particular organ or tissue. It is not uncommon for the same hormone to act in both ways. The thyroid gland is a familiar example: thyroxine has a general effect on respiratory metabolism, it will induce metamorphosis in Amphibia, and its deficiency in man leads to the defects in specific organs and tissues that characterize cretinism and myxoedema.

It is by no means easy to differentiate these two effects; for a rise in oxygen consumption or an increased rate of digestion may be the physiological consequence of some change in growth that has been activated by the hormone. The secretion from the corpus allatum certainly has some striking general effects on metabolism. When egg development in *Rhodnius* is stimulated by the corpus allatum there is a much more rapid digestion of the intestinal contents (Wigglesworth, 1936), but whether this is a direct effect of the hormone on digestion and metabolism or an indirect effect consequent upon the demands of the developing ovaries, has not been determined.

Pfeiffer (1945*a*) has suggested that the corpus allatum may influence egg production by controlling the mobilization or synthesis of materials necessary for the growth of the egg. Thus in *Melanoplus*, when the corpus allatum becomes active during the early days of adult life the insect undergoes striking changes in its metabolism. In the normal insect with the corpus allatum intact, non-fatty materials (presumably chiefly protein) are now transferred to the egg to form the yolk, and to the secretory walls of the oviducts. If the female is castrated these products accumulate in the blood and lead to retention of water and a great increase in the blood volume which distends the abdomen. If the corpus allatum is removed at an early stage, this synthesis and mobilization of protein does not take place; on the contrary, the fat body hypertrophies enormously and becomes laden with fat. Removal of the corpus allatum in *Periplaneta* likewise leads to an increase in fat from an average of 63 per cent of the dry weight of the fat body to 77 per cent (Bodenstein, 1953*c*).

Comparable changes are seen in *Calliphora* from which the neurosecretory cells have been removed: the fat body becomes stuffed with glycogen and the yolk does not develop. The eggs reach the same stage of development as in flies fed on sugar alone. It would again appear that the basic deficiency is in the production of the protein needed for yolk formation (Thomsen, 1952). Similarly, chemical analyses of *Dixippus* after removal of the corpus allatum point to a failure in the synthesis of protein, with an accumulation of amino-acids and glycogen (L'Helias, 1953*a*, *b*).

The ovaries of honey-bee workers are by no means rudimentary organs; oocytes are continuously being produced and then degenerating and being absorbed (Weyer, 1928), exactly as in the ovary of *Rhodnius* deprived of its corpus allatum. Implantation of extra corpora allata into worker bees does not induce them to produce eggs; nor are they induced to do so by the absence of the queen. But when brood cells are absent egg production begins; and for this purpose protein feeding (pollen or casein) is necessary. In other words, in the normal worker bee, protein is utilized for development of the salivary glands and feeding of the brood; only in the absence of brood is it diverted to the ovaries (Müssbichler, 1952). It is not altogether surprising, therefore, that the corpus allatum should be as large or larger in the worker bee than in the queen (Pflugfelder, 1948). Furthermore, it is at the time when the worker bee begins to collect pollen and nectar that the neurosecretory cells become most active (Scharrer and Scharrer, 1945).

In some blood-sucking insects, such as *Cimex*, *Rhodnius* and *Pediculus*, a small amount of blood is absorbed without being digested. It is then taken up from the haemolymph and deposited more or less unchanged in the yolk of the developing eggs. That suggests that much of the normal protein in the yolk may be synthesized elsewhere and merely transferred to the egg by the follicular cells (Wigglesworth, 1943). Using the methods of immunology and electrophoresis Telfer and Williams (1952) have demonstrated at least seven proteins free in the blood of female pupae of *Platysamia*, but only six in the blood of the male. The characteristic female protein, antigen 7, is about ten times more concentrated in the egg than in the blood during egg formation. But the concentration in the blood rises abnormally in castrated female pupae. Thus antigen 7 is probably synthesized by some other tissues and actively and specifically transferred to the egg. Moreover, the egg will take up the homologous blood protein of different species. In the light of earlier findings it would seem that the corpus allatum, or other components of the endocrine system such as the neurosecretory cells of the brain, may well be concerned in the production of this protein.

Another rather general effect of the corpus allatum in some insects consists in the maintenance of normal tissue growth. If the corpora allata are removed from very young *Dixippus*, in the 1st or 2nd larval stage, various degenerative changes occur in the tissues: non-fatty inclusions in the cells of the fat body, degeneration in the mesodermal sheath of the nerve cord which may cause paralysis, cystic degeneration of the Malpighian tubes and oviduct, new growths (described as 'sarcomas') in various parts of the body. These changes occur particularly in organs showing active cell divisions; they can all be prevented by reimplantation of the corpus allatum (Pflugfelder, 1938*b*). Similarly, if embryonic tissues are implanted into normal *Dixippus* they show normal differentiation; but if implanted into insects without corpora allata the upset in normal hormone production leads to abnormal tissue growths (Pflugfelder, 1947*a*). (Scharrer (1945) observed tumours in the gut and salivary reservoir of *Leucophaea* after removal of the corpora allata, but proved that this was the result of section of the recurrent nerve; they were not prevented by reimplantation of the gland.)

Finally, the combined metabolic actions of the corpus allatum may have an effect on the total oxygen consumption. Allatectomy in female *Calliphora* caused a 24 per cent decrease in the rate of oxygen uptake, and implantation of three extra corpora allata into the normal female increased the oxygen uptake by 19 per cent. In the male there was a similar fall after allatectomy, but implantation of additional glands did not cause any substantial increase (Thomsen, 1949). Extirpation of the corpora allata from young larvae of *Dixippus* caused only a temporary fall in oxygen uptake, when this was related to the body weight of the animal (Pflugfelder, 1952).

REVERSAL OF METAMORPHOSIS. It is evident that the corpus allatum exerts a wide variety of effects on metabolism. Pflugfelder (1938*a*, *b*) favoured the view that it produces a single secretion which so influences metabolism that moulting, egg formation, etc., can go forward normally. But even if, as seems probable, all these diverse effects are brought about by a single substance, whether this be termed a 'metabolic hormone' or the 'juvenile hormone', we still have to consider how this hormone controls the striking morphological changes associated with metamorphosis. This action is rendered all the more impressive by the demonstration that the changes associated with metamorphosis are to some extent reversible.

We have seen (p. 48) that the adult *Rhodnius* may be induced to moult by joining to it one or more 5th-stage larvae fed about 10 days previously. If at the same time a couple of 4th-stage larvae with the corpus allatum intact are joined on, or if six corpora allata obtained from 4th-stage larvae about a week after feeding are implanted into the abdomen, the new cuticle formed by the adult when it moults shows a partial reversion to larval characters. The change does not affect elaborately differentiated structures such as the genitalia, but it is evident in the general pattern of the abdomen (fig. 28). The distribution

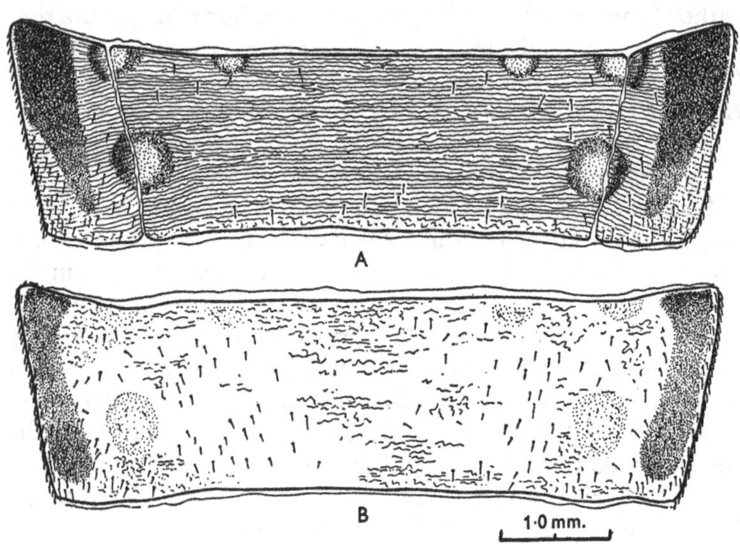

Fig. 28. A, fifth abdominal tergite of adult *Rhodnius*. B, the same segment after moulting induced by two 5th-stage larvae and the implantation of brains and corpora allata from six 4th-stage larvae. Note the increased number of bristles in B, the disappearance of the 'hinge lines', and the partial reappearance of the larval pigment spots. (Wigglesworth, 1940*b*.)

of pigment at the sides of the abdomen shows a partial return to the larval arrangement; the 'hinge line' (fig. 2) on the dorsal surface, along which the tergites fold when the abdomen is distended, disappears; the cuticle of the lateral pleat (which is highly elastic and devoid of an exocuticle in the normal adult) develops in places a pigmented exocuticle; a considerable number of new spines appear where none existed before; the form of the spines is intermediate between that in the adult and the larva, and each is often surrounded by a small plaque of larval type; and finally the new epicuticle shows the star-shaped folding characteristic of the larva (fig. 29) (Wigglesworth, 1939, 1940*b*).

Results identical in principle were obtained by Piepho (1939*a*, *b*) in the wax moth *Galleria*. When fragments of pupal integument were implanted into young larvae, and were thus induced to moult in a larval environment, they laid down a cuticle of larval type. This effect was most readily induced in that region of the epidermis which had grown outwards to restore continuity with itself ('Umwachsungshypodermis'). The epidermis immediately under the implanted fragment of pupal cuticle ('Stammbereich') often laid down cuticle of pupal type on moulting for the first time in a young larva, but produced larval cuticle if caused to moult a second or third time. In *Periplaneta*, on the other hand, if a fragment of the adult integument was caused to moult as many as four times after grafting into the cuticle of a young larva, it failed to show any reversion to the

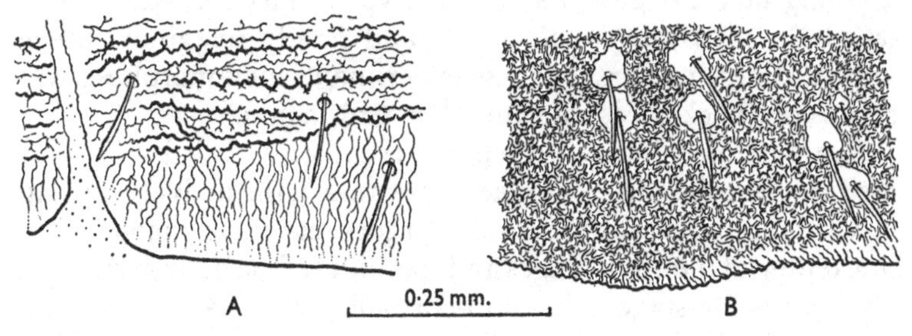

Fig. 29. A, posterior margin of fifth abdominal tergite of adult *Rhodnius*, including part of 'hinge line'. B, the same area after moulting induced by two 5th-stage larvae and the implantation of corpora allata from six 4th-stage larvae; showing partial reversion to larval type of cuticle. (Wigglesworth, 1940*b*.)

larval type, from which it differs in some rather small particulars (Bodenstein, 1953*a*).

In the case of *Rhodnius* it was observed in several instances that the larval character of the cuticle was most evident at the site of implantation of the corpora allata. This result may perhaps be comparable with that obtained by Piepho—the cells concerned in the wound-healing process being more susceptible to the action of the juvenile hormone. Or it may be due simply to this region of the cuticle being closer to the implanted glands. When small burns were made in the cuticle remote from the site of implantation they did not produce the effect.

In other experiments Piepho and Meyer (1951) and Wiedbrauck (Meyer) (1953) implanted a fragment of the integument of a full-grown larva of *Galleria* into the abdomen of another full-grown larva. It duly formed a pupal cuticle, followed by an adult cuticle, synchronously with the host. It was then removed and implanted into a young larva, into a pupating larva, or into a pupa, and finally examined in section when this second host had become adult. It was found, as in *Rhodnius*, that the imaginal integument would readily moult again to produce an adult cuticle with scales of almost normal form. In a smaller number of experiments it was possible to induce it to form pupal cuticle, and when the host moulted to become adult the implant again became of imaginal type with scales (fig. 25 C). In a very few instances, when the adult integument moulted in a young larva, it gave rise to an atypical larval cuticle which was followed by a pupal cuticle and finally imaginal cuticle with scales once more. In these experiments metamorphosis was first taken backwards and then forwards again.

Various insects undergo what is commonly called a 'hypermetamorphosis'. In Meloidae, for example, the actively feeding larva of typical caraboid form undergoes regressive changes with shortening of the appendages and rounding of the body and enters upon a resting stage or pseudo-chrysalis, often hardened like a puparium. It then emerges once more as a motile larva, now of scarabaeid form, which in course of time moults into the true pupa (Handlirsch, 1927). No information exists about the physiological control of these changes; but the parallel with the changes in metamorphosis and the experimental reversal of metamorphosis naturally suggests that hormones are again concerned.

It is instructive to consider the ability of the cells to switch over from imaginal development to larval development in the course of the moulting process. This has been studied by allowing a *Rhodnius* 5th-stage larva to commence moulting in the normal way and then, on successive days after feeding, to decapitate and provide with juvenile hormone by joining to a 4th-stage larva, about a week after feeding, in which the corpus allatum is retained.

At the temperature employed (24° C.) the interval between feeding and moulting in the 5th-stage larva averages 28 days.

Up to the fifth day the switch-over is complete; normal larval structures are laid down (fig. 30A). Thereafter the parts of the abdomen become progressively determined for the formation of adult structures (fig. 30B–G). This effect is already apparent

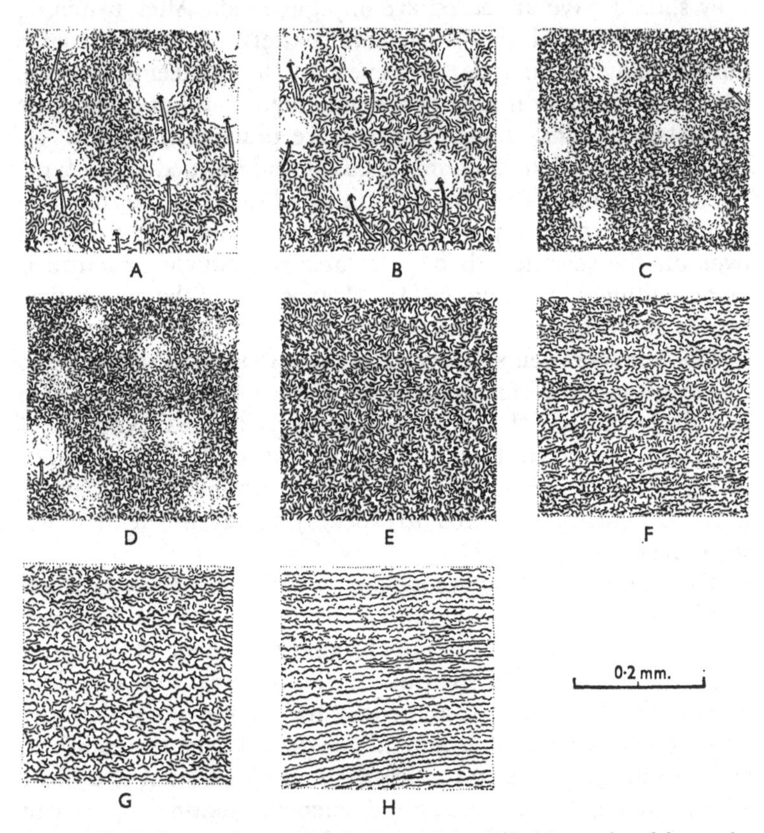

Fig. 30. Typical areas of the cuticle in the tergites of *Rhodnius* produced from 5th-stage larvae switched over from imaginal to larval development at different times after feeding. A, 5 days (normal larval type); B, 7 days; C, 9 days; D, 10 days; E, 13 days; F, 15 days; G, 16 days; H, 17 days (normal adult type). (Wigglesworth, 1940*b*.)

in the pigment pattern at the hind-end of the abdomen by the seventh day; but the plaques and bristles and cuticle over the general surface are of normal larval type. After switching over on the eighth or ninth day, the marginal pattern of the abdomen shows adult characters appearing. The plaques developed are imperfectly formed and nearly all are devoid of bristles, but they are smooth and unpigmented as usual. After operation on the tenth day very few plaques are formed, but the sites at which they should have appeared are unpigmented. After switching over on the twelfth day the general pattern of the abdomen is chiefly imaginal in type; but even on the thirteenth day the cuticle of the central area of the abdomen is still of larval type with stellate folds, though plaques are practically absent. By the fifteenth day the abdominal cuticle is intermediate in type. By the sixteenth day there are still small areas of intermediate cuticle in the central part of each segment. But after switching over on the seventeenth day or later the cuticle structure is wholly adult in type (fig. 30H). (Deposition of the new cuticle begins about the eighteenth day.)

These experiments prove that the determination of the epidermis to form imaginal cuticle proceeds at a different rate in different parts of the abdomen. Many parts remain capable of reverting to a larval type of development, under the influence of the juvenile hormone, long after the critical period; other parts are early committed irrevocably to lay down imaginal structures.

This sequence of changes is closely related with the succession of histological changes in the epidermis. Activation and cell division, with the formation of chromatin droplets as the result of cell death, take place earliest in the terminal and marginal regions of the abdomen; and these are the regions which first become irrevocably determined. The period from the eighth to the eleventh days, when the bristles and plaques fail to appear after switching over, is the period when chromatin droplets can be seen below the plaques. But it is interesting to note that, although the capacity to form the smooth cuticle of a plaque is lost by this time, the cells at these points can still lay down a cuticle that is unpigmented, as it normally is over the plaques. Clearly the various faculties of a given cell can be determined to some extent independently of one another (Wigglesworth, 1940*b*).

MODE OF ACTION OF CORPUS ALLATUM HORMONE. We are now in a position to discuss possible ways in which the juvenile hormone brings about the differentiation of adult characters. In considering prothetely in the gipsy moth *Lymantria*, Goldschmidt (1923) treated this as an example of an abnormality in morphology resulting from a disturbance in the relative velocities of two concurrent processes of development. He defined the two processes in question as 'metamorphosis' and the 'evagination of the wing discs'. He supposed that prothetely occurred when metamorphosis took place before the evagination of the wings was complete.

Goldschmidt's general conception of the control of body form by the differential velocities of the various competing developmental processes is a valuable one, which finds application in many fields of physiological genetics and developmental physiology. It may be applied to the control of metamorphosis if it be supposed that the one process is differentiation of imaginal structure and the other process is the deposition of the cuticle which brings further growth and differentiation to an end (Wigglesworth, 1936).

Thus if the 5th-stage larva of the bedbug *Cimex* is decapitated and joined to the tip of the head of the 3rd-stage larva of *Rhodnius* it is duly caused to moult; but owing to the action of the juvenile hormone from the corpus allatum of the *Rhodnius* larva its metamorphosis is inhibited. Close examination of the external genitalia shows, however, that these have differentiated some way towards the adult form. The male genitalia in *Cimex* are highly asymmetrical. This is due to the atrophy of the clasper on the right side (fig. 31 B). In the 5th-stage larva the rudiments of both claspers are visible, that on the left side being slightly larger (fig. 31 A). If the growth of the rudiments is followed during the final instar from reconstructions of the epidermal elements beneath the cuticle (Christophers and Cragg, 1922), the clasper on the left side is seen gradually to outstrip that on the right side and this finally degenerates (fig. 31 C).

Now the varied form of the genitalia in the intermediate stages produced by the action of the juvenile hormone in the 5th-stage larva agrees exactly with the form of successive stages in the growth of the epidermal rudiments (fig. 31 D). It was therefore suggested that under the action of the growth and moulting hormone the tissues begin to differentiate towards the adult form, and that the effect of the juvenile hormone is simply to speed up the process that ends in the deposition of the cuticle and in this way to inhibit metamorphosis and maintain the larval characters. It was supposed that at each larval moult a small amount of adult differentiation is allowed to take place before it is arrested by the deposition of the cuticle (Wigglesworth, 1936).

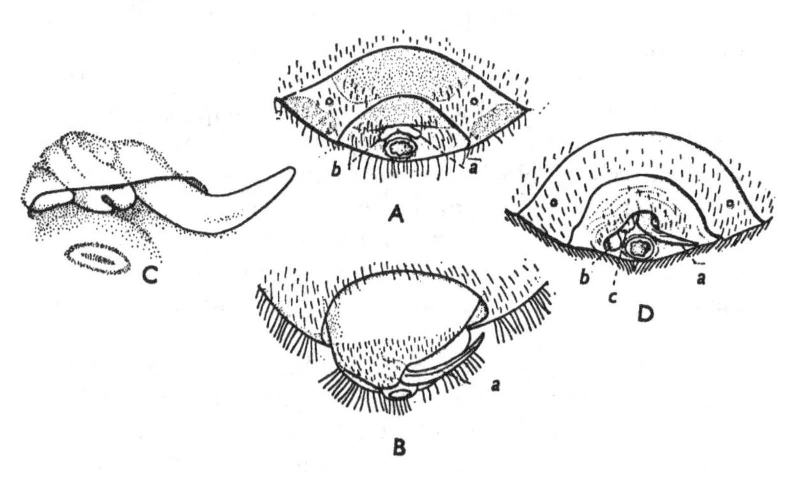

Fig. 31. Genital segments of male *Cimex*, ventral view. A, normal 5th-stage larva; B, normal adult; C, epidermal rudiments of the claspers, etc., in a male larva approaching the adult stage (after Christophers and Cragg, 1922); D, a stage with characters intermediate between 5th instar and adult, produced by parabiosis between a 5th-stage larva and a moulting 3rd-stage larva of *Rhodnius* (Wigglesworth, 1936). *a*, left clasper; *b*, right clasper; *c*, aedeagus.

It is a weakness of the Goldschmidt hypothesis that, although it provides a satisfactory description of many phenomena, it by no means follows that they are brought about in that way. Structures of intermediate form may result from developmental processes that are intermediate from the outset and not necessarily from one process overtaking another in the course of development. Indeed, when the histological changes that take place in *Rhodnius* during a larval moult and an imaginal or metamorphic moult are compared, they are found to be different from an early stage. To quote a single example: at 8–9 days after feeding, when cell division is at its height, mitoses are limited to the margins of the abdomen and the intersegmental membranes in the imaginal moult, but are exceedingly numerous all over the general surface of the segments in the larval moult (Wigglesworth, 1940*b*).

Moreover, we have seen that when the adult *Rhodnius* or the pupal integument of *Galleria* is caused to moult in the presence of the juvenile hormone a partial reversal of metamorphosis can occur, and in some regions larval cuticle can again be formed. Such a change cannot be explained by postulating two competing processes, imaginal differentiation and cuticle deposition, because imaginal differentiation is already complete at the outset. It is necessary to suppose that the dual capacity of the cell to form adult and larval characters still persists in at least some of the cells of the adult, and that the larval system can be reactivated by the juvenile hormone (Wigglesworth, 1940*b*).

QUANTITATIVE ACTION OF HORMONES CONCERNED IN GROWTH. Neither the growth and moulting hormone nor the juvenile hormone has an 'all-or-none' effect. If the moulting hormone is present in too small amounts in the full-grown larva of *Ephestia*, pupation may be only partial. It occurs in patches in the mid-line of the abdominal tergites, which appear to be regions most sensitive or reacting earliest to the hormone. And in these patches all grades of pupation can be seen, ranging from a slight hardening (exocuticle formation) in the existing larval cuticle, to a small area of fully formed pupal cuticle with detached and digested larval cuticle outside it (fig. 32) (Kühn and Piepho, 1938).

We have already noted many examples of partial metamorphosis induced by small amounts or belated application of juvenile hormone. This happens after decapitation soon after the critical period (Wigglesworth, 1934), or when imaginal development is switched over to larval development by the late introduction of juvenile hormone (Wigglesworth, 1940*b*). But perhaps the most instructive examples in *Rhodnius* are to be seen after implantation of a young corpus allatum into the abdomen of the 5th-stage larva. In the least successful experiments there may be a tiny patch of larval cuticle over the implant, which merges gradually into the normal adult cuticle around (fig. 33) (Wigglesworth, 1936). We have already noted the example in *Galleria* where there may be larval cuticle, pupal cuticle and imaginal cuticle with scales all present in the one insect (Piepho, 1950*b*).

These observations show that the juvenile hormone can have a purely local effect, and an effect which varies with the quantity of hormone available. That strongly suggests that the

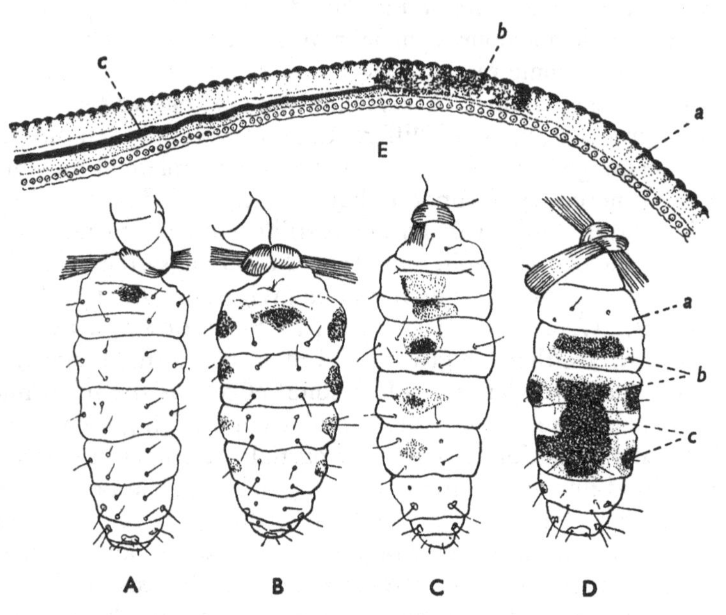

Fig. 32. A–D, abdomen of *Ephestia* larvae ligatured during the critical period showing varying degrees of partial pupation. E, transverse section through the margin of a region of partial pupation. *a*, normal larval cuticle; *b*, larval cuticle tanned and darkened but not detached from epidermis; *c*, pupal cuticle below detached larval cuticle. (After Kühn and Piepho, 1936, 1938.)

hormone is not merely catalysing or 'triggering' some cellular reaction, but is taking part in some quantitative fashion in the metabolism of the cell.

The epidermis of the insect is particularly favourable material in which to study a growth problem of this kind, because it expresses its juvenile or adult character by the type of cuticle it lays down. It is therefore often possible to say whether an individual cell is larval, adult or intermediate. The cuticle is a highly complex structure, the product of a complex system of enzymes within the cell, its characters determined by the nature of this enzyme system.

It follows that the enzyme system within the epidermal cell takes on one set of properties in the presence of the growth and moulting hormone alone, and another set of properties after combination with the juvenile hormone. Since we have no knowledge of how the living cell is organized it is impossible to say precisely how the juvenile hormone is reacting with the

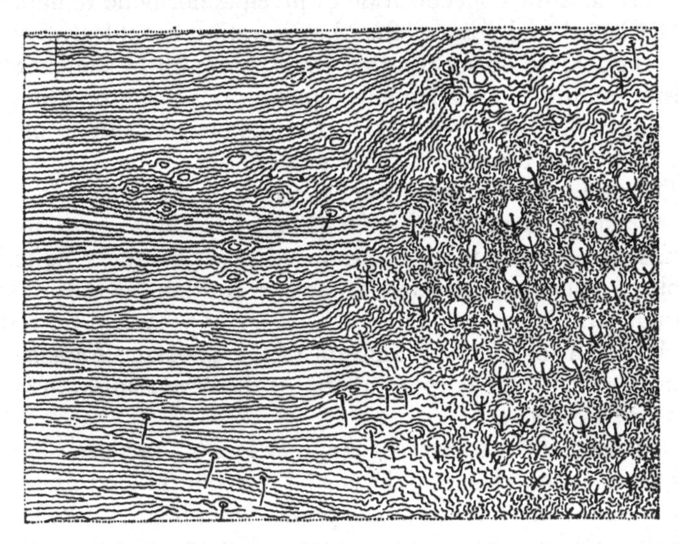

Fig. 33. Cuticle of abdomen in adult *Rhodnius*. To the right is a patch of larval cuticle with plaques overlying the site of implantation of a corpus allatum from a 4th-stage larva. Note the gradual transition in the type of cuticle, with the vestiges of sockets without bristles in the intermediate zone. (Wigglesworth, 1936.)

enzyme system of the cell. But it may be well to recall that not only is the same hormone effective in causing development of larval characters in quite unrelated insects, but it has totally different effects in different parts of the body. To recall one simple example: it causes cells in one part to lay down colourless cuticle in the larva while they form black cuticle in the adult, and cells of another part to form black cuticle in the larva while they form colourless cuticle in the adult (p. 53). It must, therefore, be acting in some very general way.

In spite of the fact that cuticle or single bristles intermediate between those of the larva and adult may be produced in *Rhodnius*, the fact remains that intermediates are on the whole uncommon. In Lepidoptera, the cuticle or the patches of cuticle laid down in the experimental animals are nearly always either larval, or pupal, or imaginal. When the corpora allata are excised from a caterpillar, no matter what instar it may be, it always forms the pupa, which develops into the imago. It seems unlikely that the concentration of juvenile hormone remaining in the blood and tissues after the operation should always be the same. It would seem rather that there must be some mechanism tending to canalize development along one or other of three lines; or, in other words, that the enzyme system of the cell must have three alternative phases.

The same surprising constancy is seen in the morphology of the 6th-stage larva of *Rhodnius* (Plate I*d*). This instar never occurs naturally, but when produced experimentally its morphology is just as characteristic and as constant as that of the normal 5th-instar—no matter whether it is produced from the implantation of a corpus allatum of the 4th-stage *Rhodnius*, a number of 1st-stage corpora allata, or the corpus allatum from an adult *Periplaneta*.

CONTROL OF FUNCTION OF CORPUS ALLATUM. Metamorphosis is controlled by the corpus allatum. When this ceases to secrete the juvenile hormone, or secretes it in greatly reduced amount, metamorphosis takes place. But the question arises, how is the activity of the corpus allatum controlled?

If the corpus allatum of a 4th-stage larva of *Rhodnius* is implanted into the abdomen of a 5th-stage larva, the latter develops into a 6th-stage larva, and the implanted corpus allatum is now in the 5th-stage—which does not secrete the juvenile hormone. When these 6th-stage larvae moult again, some of them give rise to adults, but some of them give rise to 7th-stage larvae in which very little further differentiation of adult characters has taken place. Clearly the implanted corpus allatum is still producing juvenile hormone (Wigglesworth, 1936). In other experiments of this type (Wigglesworth, 1948*a*) the results are equally variable, and it is clear that when the insect undergoes more than one moult after the implantation has been made, development of larval or adult characters bears no strict relation to the stage from which the corpus allatum was derived.

Clearly the corpus allatum cannot be wholly autonomous: it does not, for instance, produce juvenile hormone for a definite number of moulting stages. In other words, it is not the corpus allatum itself which 'counts the instars'. The mechanism responsible for this presumably resides in the central nervous system, and the requisite stimulus must be conveyed to the corpus allatum by means of the nerves with which it is richly supplied.

The central nervous system is presumably responsible therefore for the number of moulting stages passed through before metamorphosis occurs. This number may be influenced by environmental conditions. In the silkworm the number of larval stages varies from three to five. This number is influenced by the temperature and the length of day to which the young larva or even the eggs during incubation are exposed (Kogure, 1933). *Blattella* normally has six moults, but when given a less nutritious diet or when subjected to minor injuries, such as amputations of limbs, it makes seven moults, the growth ratio (p. 19) at each moult being correspondingly reduced (Seamans and Woodruff, 1939). A raised temperature increases the number of moults from nine to eleven in *Sphodromantis* (Przibram and Megušar, 1912), from four to five and from five to six in males and females respectively of *Dermestes* (Kreyenberg, 1928), from an average of four at 18° C. to an average of five at 25° C. in *Ephestia*. It has the reverse effect in *Pieris* which moults five times at 14–15° C., four times at 15–20° C. and three times at 22–27° C. (Klein, 1932).

The number of instars may be hereditary; for example, different races of *Bombyx mori* have the number of ecdyses determined by Mendelian factors (Ogura, 1933). Key (1936) described a strain of *Locusta migratoria* in which all the females had an extra (6th) instar irrespective of the conditions of culture, whereas all the males had the normal five instars. This extra instar might be an extra 3rd or an extra 4th. That is, even within the strain the course of reduction in the activity of the corpus allatum could vary slightly. Such occasional abnormalities serve to emphasize how accurately the process of morphological change is normally regulated.

CHAPTER 5

DIFFERENTIATION AND POLYMORPHISM

We have now formulated a tolerably consistent theory of the control of metamorphosis. The insect is a dimorphic or, if there is a well-defined pupal stage, a trimorphic organism. Its growth is initiated by a hormone secreted in the thoracic gland (including in that term the ventral, pericardial, prothoracic and peritracheal glands, all of which are believed to be homologous or at least analogous). But this gland must first be activated by a factor (called by Scharrer (1952*b*) the 'prothoracotropic hormone') produced in the neurosecretory cells of the pars intercerebralis (or perhaps elsewhere in the brain (Pflugfelder, 1949)). In the presence of the thoracic gland hormone, alone or when it greatly predominates, the imaginal characters are differentiated in the growing tissues. If the juvenile hormone, secreted by the corpus allatum, is present in addition in high concentration, the larval characters are differentiated; and, in consequence, the latent imaginal characters are suppressed, so that metamorphosis may be said to have been inhibited. Progressive reduction in the amount of the juvenile hormone secreted, or slight changes in the timing of secretion, may result in very small instalments of imaginal differentiation throughout the larval stages of hemimetabolous insects. A substantial reduction in the amount secreted in holometabola results in the development of pupal characters. A greater reduction, or sometimes perhaps a complete absence of juvenile hormone, leads to the formation of the imago. But in many organs a certain minimum amount of growth is necessary before pupal or imaginal differentiation is possible; such organs must reach a certain size before they become 'competent' to undergo metamorphosis.

In the present monograph the hormone of the thoracic gland has been termed the 'growth and moulting' hormone, because each cycle of growth which it initiates terminates in the moulting of the old cuticle. It is called by Scharrer (1946*a*) the 'growth and differentiation hormone'; but this term has been avoided for two reasons: (i) As will be made clear in this chapter, differentiation among growing cells commonly results from their mutual interaction in a manner which cannot be ascribed to the activity of the growth and moulting hormone. (ii) The use of the term 'differentiation hormone' implies that the insect larva becomes progressively differentiated to form the adult; that the insect larva is indeed an embryo (p. 1). But in many insects the organs of the larva are as highly differentiated and as specialized in their own way as those of the adult.

These hormones serve merely to release, control and direct the inherent capacity for growth and differentiation possessed by the cells of the various tissues. By themselves they can do little to ensure the proper integration of growth in the different parts of the body; and the true nature of the changes in form which they bring about cannot be understood until we know something of the 'substrate', the living substance of the body on which they act. These are the problems with which we shall be concerned in this final chapter.

INTEGRATION OF GROWTH. One of the most impressive features of normal growth is the synchronization and coordination of the process in all parts of the body. It is clear that the secretion of hormones into the circulating blood will ensure that growth begins everywhere at the same time, but they can hardly ensure that growth remains coordinated throughout a moulting period which may vary greatly in length from one individual to another.

It may be that there is a series of hormonal signals given out as one phase of growth succeeds another. This might provide a function for the neurosecretory cells which occur throughout the thoracic and abdominal ganglia in insects. At the final moment of moulting there is a rapid absorption of the moulting fluid beneath the cuticle and in the tracheal system; the new cuticle is waterproofed by the secretion of wax; then the skin is shed, the 'cement' is discharged from the dermal glands, and hardening and darkening of the cuticle begins. When the mosquito larva *Aedes* hatches from the egg there is the same absorption of fluid from the tracheal system. The temporary suppression of absorption by narcotizing with chloroform suggests that the central nervous system is controlling this final growth change; and since the tracheal cells are not innervated it seems likely that some humoral control is involved (Wigglesworth, 1938). The same applies to the hardening and darkening of the cuticle in the adult *Calliphora*, which normally occurs within 1–2 hours of emergence from the puparium, but may be deferred for 7 hours or so if the fly is obliged to keep digging its way through the soil (Fraenkel, 1935*b*).

There is no experimental evidence to prove the existence of these hormonal signals; for the present they are hypothetical. But it is interesting to note that in Ephemeroptera and Odonata there is a sudden discharge of secretion from the neurosecretory cells along the axons just at the moment of ecdysis. Arvy and Gabe (1953*b*) compare this with the rapid discharge of secretion from the cells of the hypothalamus that is seen in the rat subjected to emotional shock.

Another possible medium for maintaining integrated growth is afforded by the cellular continuity of the epidermis. It was noted by Crampton (1899) that if pupae of the large silkmoths were joined together with paraffin wax after removing small areas of the integument (fig. 9A), the epidermis became continuous from one to the other, and emergence always took place simultaneously in the two insects. The same occurs in *Rhodnius* where epithelial continuity may be established with insects of other genera or other families, such as *Triatoma* or *Cimex* (fig. 6). Here again the united insects moult simultaneously; whereas it was observed that if they were connected by a capillary tube they very often moulted independently of one another (Wigglesworth, 1936). Bodenstein (1938) likewise noted that joined fragments of *Phryganidia* pupae show simultaneous development if an epithelial union is present, as also do *Periplaneta* in parabiosis (Bodenstein, 1953*a*).

These observations suggest that synchronous development in different parts of the integument may perhaps be maintained by chemical or other stimuli transmitted by way of the continuous epidermis. But the problem requires much more critical study before this can be regarded as proved.

RELATIVE GROWTH OF PARTS OF THE BODY. The varying degrees of partial pupation that occur in caterpillars exposed to small amounts of the moulting hormone (p. 91) show (i) that the different parts of the body have different thresholds of response to this hormone and (ii) that the hormone has a quantitative effect on growth and not merely a catalytic or 'triggering' effect. Presumably a structure which grows more during larval life is consuming more growth hormone than a structure which grows less. The wing disc in Lepidoptera will be expected to absorb more hormone than a leg disc which does not appear until a late larval stage.

Certainly the different parts of the body have different capacities for growth. That is the chief factor responsible for the characteristic shape of the organism. It may well be that local differences in growth activity are correlated with differences in the absorption of hormone, and that the visible form of the organism is preceded by an invisible pattern of 'growth capacity'.

But it is not the supply of hormone alone which limits the amount of growth. There are many other nutritional factors which may set a limit to the amount of growth that can be achieved in a given period. All these factors are equally available to all parts of the body. Each part will get its share in accordance with its avidity in absorption; that is, in proportion to its capacity for growth.

It follows, therefore, that if any one of these factors sets a limit to the amount of growth, it will have an integrated or proportionate effect upon all parts of the body. It is this proportionate growth of the various parts which usually follows the allometry law (p. 19).

In seeking an explanation for the outgrowths of enlarged structures characteristic of the adult insect, Novák (1951 *a*, *b*) ascribes these to the existence of an intracellular growth factor localized in the regions where such outgrowths occur. This substance is termed a 'gradient factor', which is supposed to diffuse from cell to cell and to enable the parts of the body in which it occurs to grow in the absence of the juvenile hormone. At the present time this is a purely hypothetical way of accounting for the fact that at the time of metamorphosis, and to a lesser extent in each larval stage, certain regions of the epidermis do show excessive growth activity. The 'gradient factor' does not differ from the 'latent imaginal system' which we have been postulating throughout this book, except that it is supposed to be in some degree diffusible and in his latest writings on this subject Novák (1954) describes the 'gradient factor' as being bound to the cell plasma and no longer conceives it as being diffusible.

In the discussion above, local differences in growth have been attributed, in non-committal terms, to local differences in growth capacity. But in the vas deferens of *Drosophila* there is evidence of something of the kind suggested by Novák (Stern, 1941). There are striking differences in the testes of different species; *D. virilis* and *D. azteca* normally have coiled testes, in *D. pseudo-obscura* the testes are uncoiled. Transplantation experiments show that these differences in the degree of coiling of the testis are determined by the vas deferens with which it is in contact. Stern suggests that the vas deferens may do this by releasing a growth-promoting substance which diffuses by direct contact into the growth region of the attached testis. If this substance, acting like auxin in plants, is given off in different amounts to opposite sides of the testis, it will induce different degrees of growth with consequent spiralization.

NATURE OF THE 'SUBSTRATE'. It is evident that the hormones concerned in growth and metamorphosis are doing no more than regulating and controlling the realization of the potentialities latent in the cells. It is therefore impossible to understand the nature of metamorphosis until we understand the nature of the substrate upon which the hormones act; that is, until we understand the processes of determination and differentiation and the role of the genes in controlling these changes.

To discuss these matters fully would require a review of the whole subject of growth and development. All that will be attempted here is to bring forward certain views and ideas that furnish a provisional framework of hypothesis upon which the established facts can be held together and displayed. This is worth doing in the present context because the information we now have about the control of growth and metamorphosis in insects can be made to play a substantial part in the argument; for the study of the reaction between the growth hormones and the tissue cells may itself throw light on the nature of the substrate.

The earliest stage of the organism is the young oocyte in the ovary, which is nourished in the follicle to produce the mature egg. Now the egg in the follicle is part of the phenotype of the mother; it is always orientated in the same axis as the mother. It is difficult to conceive the chromosomes in the nucleus, which are presumably disorientated, being responsible for inducing this polarity in the egg. It seems more probable that the oocytes are orientated from the outset. They have either retained the polarization they received as 'pole cells' in the egg from which the mother developed, or the polarity of the egg has been induced in it by the follicle in which it lies, just as the follicle must certainly determine the form and orientation of the shell. In either case the polarity of the egg, the earliest character we can recognize, is a property of the cytoplasm.

The importance of the cytoplasm is apparent throughout growth and development. We have seen that in the egg of *Tenebrio* the germ band zone may be visible in the cortical plasma of the egg before cleavage begins (Ewest, 1937), and that in the egg of *Drosophila* and muscid flies at the time of laying, the cortical plasma is a mosaic of zones already determined for the formation of particular regions of the larva. That is not to say that the nuclear genes are not concerned in this process. There can be no doubt that they have been vitally concerned at an earlier stage in influencing the changes in the egg cell. And defects in the chromosomes (such as deficiencies in the X-chromosome in *Drosophila* (Poulson, 1945)) can lead to disruption of development in the early stages of cleavage so that no blastoderm is produced and all subsequent development is lacking. But that does not alter the fact that the regional diversity in the egg, which is the precursor of differentiation in the organism, is a property of the cytoplasm. That we shall find to be so throughout development.

It is, indeed, generally believed that the genetic constitution of the nuclei is uniform throughout the organism and that the diversity which results in differentiation resides in the cytoplasm; 'that the differences in cell heredity that arise within a multicellular organism are cytoplasmic' (Wright, 1945). It was to provide a physical basis for this diversity that 'plasmagenes' have been postulated. These are pictured as being controlled or synthesized by the nuclear genes and transmitted through the cytoplasm of the egg. The plasmagene is conceived as a self-duplicating material within the cell which can become modified chemically and multiply and reproduce itself in its modified state. It is further supposed that the chemical modifier or 'hapten' may be the product of an intracellular gene-controlled chain of reactions or an inductor or hormone acting from without (Wright, 1945).

The demonstrable properties of a cell that has become differentiated are to be measured only by the character and coordination of the enzyme systems it contains. In the last analysis a cell that has become differentiated is a cell in which the enzyme system has undergone some change. In recent years there has been an increasing tendency to regard the intracellular enzymes as being built into the structure of the cell. Edlbacher (1946), for example, suggested that in the living system enzymes are organized and combined to form polyvalent enzyme complexes which are thus able to carry out virtually simultaneously the whole sequence of chemical changes that the substrate undergoes, thus making possible the enormous velocity of many metabolic reactions. Experimental evidence proves that that, in effect, is what the mitochondria are.

So far as the cell as a unit is concerned, the idea of a 'cyto-skeleton' occurs again and again in the literature. The idea was very clearly expressed by Baitsell (1940), who goes so far as to suggest that 'the cell is essentially a protoplasmic crystal in which an almost infinite number of protein molecules, beginning with the genes in the chromosomes, are associated in a definite ultra-microscopic pattern characteristic of the particular type of cell'. Individual protein molecules would not be present in such a system; all would be united into a continuous substance. It may be that in their association to form the multicellular organism, the cells 'are only repeating the condition of the protein molecules which are bound together in the protoplasm of each cell to make the fundamental cellular unit' (Baitsell, 1940). Lillie (1909, 1929), too, was inclined to seek 'a molecular basis for the fundamental principle of vital organization'; Edlbacher (1946) has suggested that the developmental potencies of the living system are to be sought in the molecular structure of a species-specific matrix; and Costello (1948) visualized the unfertilized egg as a bilateral liquid crystal structure extending throughout the cell, serving as the 'framework' of the cytoplasm.

In earlier publications (Wigglesworth, 1945, 1948*a*) the idea was developed that the supracellular fabric of the organism is indeed a chemical continuum, a 'molecule' in the sense that it is held together by chemical bonds, and that it is the continuity of this substance which provides for the unity of the organism. In its primordial state, as it exists in the cortical plasma of the germ-band zone of the egg, this species-specific matrix requires only to be activated in the appropriate manner to give rise to all the enzyme systems which characterize the different parts of the organism and whose activity defines its form.

This substrate (which must already have undergone some minimal amount of differentiation, since it is polarized) is now pictured as reacting with some substance, some 'modifier' or 'inductor' which renders a part of it capable of some new activity. That is 'determination'. Later the modified zone becomes active in this new way and that leads to visible 'differentiation'.

Why is the determination of a particular function, say the determination of the precursor region of the head, limited to a particular part of the substrate? One may suppose that the modifier is limited in amount. It is taken up by, or combines with a particular part of the substrate, perhaps because that region occupies the highest level in some gradient along the axis of the polarized egg. Once that region has been activated in this way all other regions will be inhibited because the available supplies of the modifier are drained into that region. Another modifier (or perhaps the same modifier at a lower concentration) becomes available and another region, say the thorax, combines with it and so becomes appropriately determined and in its turn inhibits like determination elsewhere. If the germ is divided before this sequence of modifications has begun, the two halves will undergo just the same changes, and duplication will result. The process is somewhat analogous to the periodic precipitation which gives rise to the Liesegang phenomenon. Here also the zones of precipitation are surrounded by zones in which precipitation has been inhibited by the removal of one reactant.

According to this conception the specific characters will be determined by the chemical nature of the basic substance and the nature and quantities of the modifiers with which its parts interact. These parts may lie in a non-cellular continuum, or they may be subdivided into cells. That will not affect the general conception. They will be equally subject to the products of the nuclear genes. These may be assumed to be identical in all the nuclei and therefore in all the cells. But they will be able to exert their effects only upon those parts in which the cytoplasmic substrate has the appropriate properties to react.

Of course, when the animal or the part exceeds a certain size and complexity it is provided with a skeleton and with mechanisms of coordination of quite other kinds. It can then no longer be properly described as a 'giant molecule'. But even in large and complex organisms the same principles may be conceived as operating in the local points of growth and regeneration.

Certain parts of this conception have been developed by Goldschmidt (1927, 1938); others will be found in the writings of Jensen (1948) and Weiss (1950, 1953); the idea of a molecular basis for differentiation has been applied by Fauré-Fremiet (1948) to explain morphogenesis in ciliates. It has been set out here in brief dogmatic form because it serves very well to provide a consistent explanation for many of the observations of growth in insects. Indeed, some parts of the conception were arrived at in the attempt to explain the results of experiments on the growing insect (Wigglesworth, 1940*a*, 1945, 1948*c*).

DETERMINATION IN EPIDERMIS OF *RHODNIUS*. It must be supposed that when a cell carrying a part of the modified substance in its cytoplasm divides, the modified component, duly increased in amount, is shared between the daughter cells in the same kind of way as the nuclear substance. That is why the term 'plasmagene' is often applied to this cytoplasmic constituent. Indeed, pursuing the analogy still further, Wright (1945) describes differentiation as 'controlled mutation of plasmagenes'—although, as Ephrussi (1953) points out, the increase of some cytoplasmic constituent does not necessarily mean that it is self-reproducing in the manner of the chromosomal genes: it may grow by accretion like a crystal.

One of the most familiar examples of a modification of this

kind is the 'transformation principle' (apparently a desoxyribose nucleic acid) which will induce a permanent inheritable transformation of unencapsulated variants of Type II pneumococci into the fully encapsulated type (Avery, MacLeod and McCarty, 1944). We have seen what looks like a similar phenomenon strikingly displayed in the enlargement and displacement of the pigment spots on the abdomen of *Rhodnius* following a burn (p. 52), when the cells differentiated to carry out the enzymic processes needed for laying down black cuticle multiply and spread over the wound.

At the stage at which they have been studied the cells responsible for the formation of black cuticle have already been determined. But over the general surface of the abdomen in the larval stages of *Rhodnius* there are smooth rounded plaques each bearing an innervated bristle, and there are small dermal glands. Both these structures increase in number from one instar to the next. They arise by differentiation from the ordinary epidermal cells. They therefore provide excellent material in which determination and differentiation can be studied experimentally.

The formation of these plaques is not predetermined at an early stage; they are induced to differentiate in the course of development. New bristles and plaques are added at each larval moult, and the number added is not constant. That is readily shown by the fact that the number of plaques can be varied experimentally. The extreme case is seen after the healing of a burn, when they are regenerated at regular intervals over the wide area that has been repaired (fig. 34).

Close examination of the integument before and after moulting shows that over a given area it is the separation of existing plaques which leads to the emergence of new ones. The new ones make their appearance where the existing gaps are most obvious (fig. 35). By blocking the anus with paraffin immediately after a large distending meal of blood, it is possible to increase abnormally the separation of the plaques; and then an increased number of new plaques make their appearance—not at the moult which follows immediately, but at the moult after that (Table 1).

By decapitating a larva that has not been fed and then transfusing it with the blood of a larva in process of moulting, it can be caused to moult. In such an insect, which has not been distended by a meal, the plaques remain close together; yet there is the same increase in the number of plaques as in the insect which has had a normal meal (Table 1). And if a similar insect

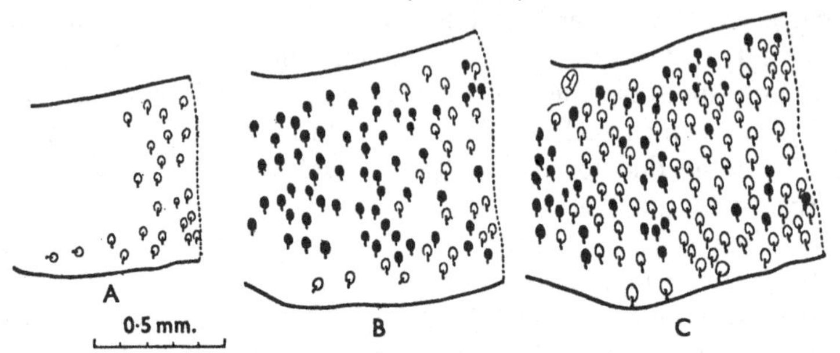

Fig. 34. Regeneration of plaques in *Rhodnius* after burning. A, part of abdominal segment of 3rd-stage larva after a burn in the 2nd-stage; plaques are absent over the healed area. B, the same in 4th-stage larva; plaques developed at site of previous plaques shown in outline, newly differentiated plaques black. C, the same in 5th-stage larva. (Wigglesworth, 1940*a*.)

TABLE 1. *Effect of distension of the abdomen on the formation of new plaques during moulting of the 4th-stage larva*

Experimental treatment	Percentage increase in the number of plaques on the abdominal tergites
4th stage, normal	27·2
4th stage, excessively distended by blockage of the anus	26·7
4th stage, unfed and undistended (moulting artificially induced by transfusion)	27·4
4th stage, normal—following excessive distension in third stage	43·8

has been subjected to excessive distension at the preceding moult, there is the usual *excessive* increase in the new plaques formed.

It might be argued that the new plaques had already been determined even before the meal of blood is given. But that cannot be so; for we have seen (p. 65) that the number of new plaques formed at the moulting of the 4th-stage larva (27·5 per cent increase) is less than at the moulting of the 3rd-stage larva (50 per cent increase); and if the 4th-stage larva is decapitated and joined to a 3rd-stage with its corpus allatum intact, the percentage increase in the number of new plaques is raised to an average of 36·3 per cent. Clearly the plaques are not predetermined at the commencement of the moult, but are probably determined at an early stage in the moulting process.

During moulting, as we have seen (pp. 12, 20), after a period of exuberant mitosis with the breakdown of excess nuclei, the epidermal cells settle down with the nuclei evenly dispersed at a constant distance from one another. The number of cells per

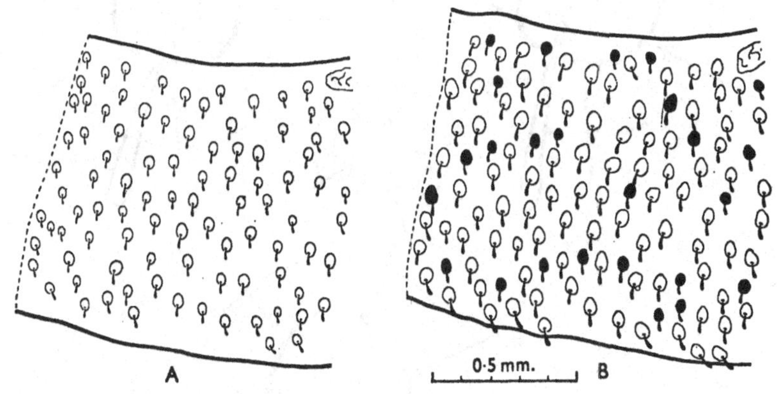

Fig. 35. Formation of new plaques in normal *Rhodnius*. A, part of abdominal segment of 4th-stage larva; B, the same after moulting to 5th-stage; newly differentiated plaques black. (Wigglesworth, 1940*a*.)

unit area is therefore constant. The plaques, on comparable areas of the body surface, are therefore separated by a constant number of cells. At the commencement of moulting this number will be the same whether the insect has been distended with a meal or remains unfed. It would seem, therefore, that the new plaques appear at a more or less fixed distance from existing plaques—but this distance is measured, not by the absolute interval of separation, but by the number of epidermal cells intervening.

This can only mean that the realization of the potentialities of the cells to form new plaques is controlled by the existing plaques. Each plaque appears to inhibit the emergence of new plaques within a certain distance of itself. What the nature of this inhibition may be we do not know, but in accordance with the hypothesis outlined above, we may think of the plaque-forming centre as absorbing from the surrounding cells some material essential for the determination of the plaques and so

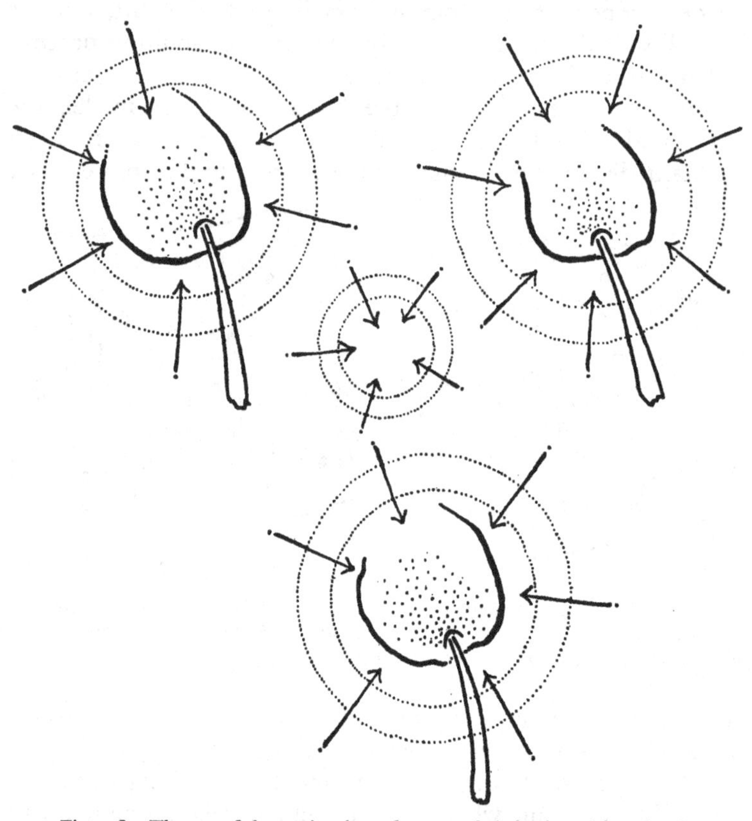

Fig. 36. Theory of determination of sensory bristles in epidermis of *Rhodnius*. Explanation in text. (Wigglesworth, 1940*a*, 1953*b*.)

draining the adjacent area and rendering the cells incapable of determination for this purpose. Beyond a certain distance, as measured by the number of cells intervening, the essential substance remains above the threshold level. A given cell in this area appropriates this substance, drains it from the surrounding zone and becomes a new centre for plaque formation (fig. 36).

The process is analogous to that which determines the emergence of universities and many other institutions in human communities (Wigglesworth, 1940*a*, 1945, 1948*c*).

But the situation is more complex than this; for besides the plaques there are the dermal glands. And in the earliest stages of their development the dermal glands, like the plaque-forming centres, consist of four small nuclei. Indeed, at this early stage the two structures are indistinguishable. The dermal glands are

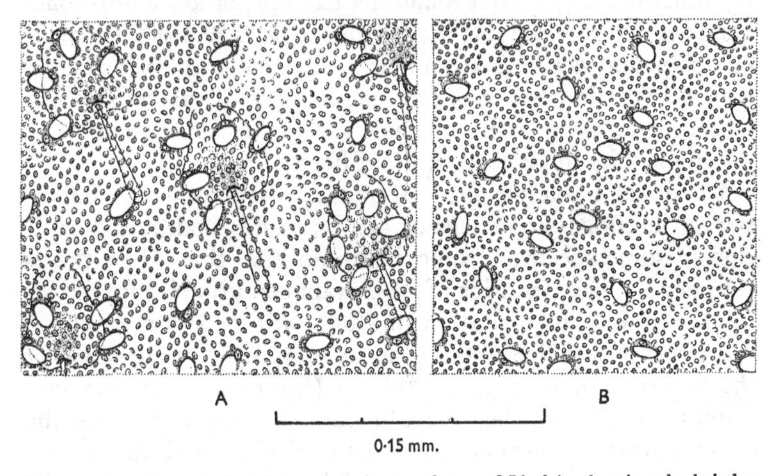

Fig. 37. A, epidermis of abdomen in 4th-stage larva of *Rhodnius* showing the bristles and plaques, and the distended dermal glands. B, the same at the first moult after a burn; dermal glands have been differentiated in the regenerated epidermis but not bristles. (Wigglesworth, 1953*a*.)

more numerous than the plaques; they occur in groups of four or five around each plaque with occasional glands in the spaces between (fig. 37A).

Now at the first moult after the repair of a burn no new plaques or bristles are developed. But new dermal glands appear, and are distributed evenly over the healed area (fig. 37B). The dermal glands arise much closer together than the bristles. There is some evidence that they are homologous structures. It may well be that the same substance is necessary for the determination of both, but that it is needed at a higher concentration or in greater amount for the determination of a plaque-forming centre than for a dermal gland. That might explain why the glands appear in the newly formed epithelium after burning whereas the plaques do not arise until the next moult (Wigglesworth, 1953*a*).

If the adult *Rhodnius* is caused to moult, the dermal glands often give rise to finger-like outgrowths which might be mistaken for aberrant bristles. Apart from this, intermediate stages between glands and bristles are not seen in *Rhodnius* (although they are described as occurring commonly in larvae of *Tenebrio* (Plotnikov, 1904)). That is another example of an 'all-or-none' response in development; there seems to be some mechanism for ensuring that determination goes either one way or the other. As a schematic possibility one might think of the cytoplasmic substrate reacting with one molecule of the inductor or with two molecules to give two quite different results.

Chemical inductors are not readily demonstrated; but in various mutants the deficiency can be made good by the addition of some specific chemical substance so that when this is supplied the manifestation of the mutant condition is suppressed. Kynurenine or its oxidation products will lead to normal eye coloration if supplied to the *vermilion* mutant (*v*) of *Drosophila* or the red-eyed mutant (*a*) of *Ephestia* (Ephrussi, 1942); various iminazol derivatives have an 'anti-bar' effect and cause the development of normal eyes when administered to *bar* eye *Drosophila* in which the number of facets would otherwise be greatly reduced (Chevais, 1943; Butenandt, Karlson and Hannes, 1946); the provision of riboflavine in the diet of *Drosophila* of the mutant *antennaless* results in the development of normal antennae (Gordon and Sang, 1941). These effects may not be precisely comparable with the action of inductors, but they serve to illustrate the striking and specific morphological changes that can be brought about by a single chemical.

Morphogenesis in the insect is thus primarily a question of self-regulation within the epidermis. The epidermis itself controls the realization or the suppression of the potentialities which reside within its cells. This process of inhibition or release of morphogenetic potencies can go forward only in a medium in which there is continuity. It is therefore to be expected that when a part of the organism is divided before determination is complete, the same process will take place in the two halves.

This may be seen in the embryo itself (Seidel, 1936; Krause, 1953), in the wing anlagen of Lepidoptera (Magnussen, 1933), in regenerating limbs (Bodenstein, 1937) or in the ocellar spots or other elements in the wing pattern of Lepidoptera (Henke, 1933). The latent imaginal organism may be capable of regulation until quite an advanced stage: X-ray treatment of the early 3rd-stage larva of *Drosophila* may result in doubling of palps,

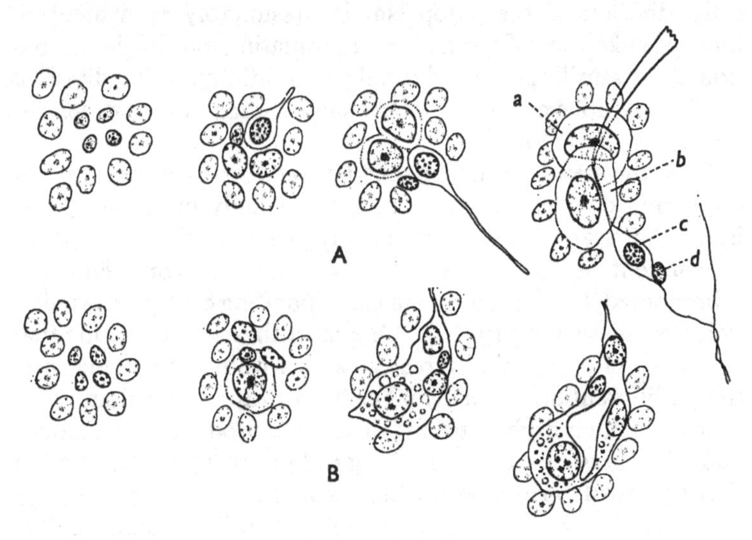

Fig. 38. A, stages in the differentiation of sensory bristle in integument of *Rhodnius*; B, stages in the differentiation of dermal gland. *a*, tormogen cell; *b*, trichogen cell; *c*, sense cell; *d*, neurilemma cell. (Wigglesworth, 1953*a*, *b*.)

scutellum and wings (Waddington, 1942). The male genital disc of the *Drosophila* larva is already divided mosaic-wise into fields, but the parts of each field are capable of regulation or reduplication if divided experimentally (Hadorn, Bertani and Gallera, 1948).

DIFFERENTIATION IN EPIDERMIS OF *RHODNIUS*. At some time after the centres of formation of the sensory hairs or of the glands have been determined, visible differentiation takes place (fig. 38). In both structures it seems that determination has affected initially one cell only. This cell divides to give four apparently identical cells with small nuclei. In the case of the sensillum, which forms the centre of the future plaque, the four cells become respectively the tormogen or socket-forming cell, the trichogen cell, the sense cell and the neurilemma cell (fig. 38A). In the case of the dermal gland one cell becomes a large glandular cell, there are two cells with smaller nuclei, and one cell with a very small nucleus resembling the neurilemma cell (fig. 38B).

In the growth and multiplication of the ordinary epidermal cells, division of the cytoplasm is presumably equivalent or meristic, like that of the nuclear chromatin. But in the formation of the sensillum or the dermal gland differentiating divisions occur; the differing capacities are shared unequally between the four daughter cells.

Now the new sensilla arising between existing plaques, or appearing over a healed burn, are normally orientated; that is, the hairs are directed posteriorly; whereas if a piece of the integument is excised, rotated through 90° or 180° and reimplanted, the bristles show a corresponding change in orientation (Wigglesworth, 1940*a*). One must therefore conclude that some sort of orientation already exists within the cytoplasm of the undifferentiated epidermal cells; with the result that, when they make their differentiating division to produce the tormogen and trichogen cells, these are so placed in respect to one another that the bristle grows out in the predetermined direction.

These observations point to the existence of some kind of 'cytoskeleton' within the undetermined cell, which defines the antero-posterior axis and controls the mutual relations of the daughter cells (themselves perhaps forming a syncytium) and in this way controls the orientation of the resultant structures —just as a unicellular protozoon may have a highly differentiated structure.

The sense cell which is one of the four daughter cells arising in this way is a new sensory neurone. Such neurones are thus differentiated from the ordinary ectodermal cells throughout post-embryonic life. The inwardly growing axons from these cells illustrate another common phenomenon in the relations between differentiated cells, the mutual affinity of the same cell types. The new axon may grow in any direction, between the epidermal cells and the basement membrane, until it makes contact with an existing nerve or axon. It then proceeds to accompany this nerve, which it follows through the basement membrane and so to the central nervous system (Wigglesworth, 1953*a*).

These four cells probably represent 'terminal forms' which are not capable of undergoing further differentiating divisions. But their functions are still susceptible to extensive modification in other ways.

They may be influenced by the genetic constitution of the nucleus. This is evident in *Drosophila* in which there is a great variety of genetically determined bristle characters which arise from disturbances in the timing or the quality of the secretion in the trichogen cell (Lees and Picken, 1945). And they may be influenced by the hormones controlling metamorphosis. If 3rd-stage larvae of *Rhodnius* are decapitated around the critical period, bristles showing all intermediate stages between the normal larval type and the adult type are developed in different specimens, depending upon the quantity of juvenile hormone present in the blood at the time of decapitation (fig. 39).

HOMŒOSIS. The epidermis of the abdomen in *Rhodnius* has been considered in some detail because it provides a relatively simple model of determination and differentiation in a system where the cellular changes can be followed step by step. Differentiation of sensilla in the epidermis, or their regeneration over a burned area, are the same processes and are presumably controlled by the same mechanisms, as organ formation and regeneration on a larger scale. The diversity of structures formed by the epidermis of the abdomen is an example of polymorphism, in every way comparable with the polymorphism of different organs of the body or different individuals of a species.

Homœosis is the name given to the appearance in one segment of the body of appendages or other structures which properly belong to another segment. Here again we have a relatively simple abnormality of differentiation which may be expected to throw light upon the normal process. According to the theory of differentiation developed by Goldschmidt (1927, 1938), the local effects are produced by general changes impinging upon growth processes which are proceeding at different rates in different parts. To that extent the purely localized action of genes is illusory—the genes are merely affecting the relative velocity of the different processes concerned.

This explanation was applied by Goldschmidt (1927) to the differentiated wing patterns of Lepidoptera. Put in its most schematic form the hypothesis supposes that the wing scales go

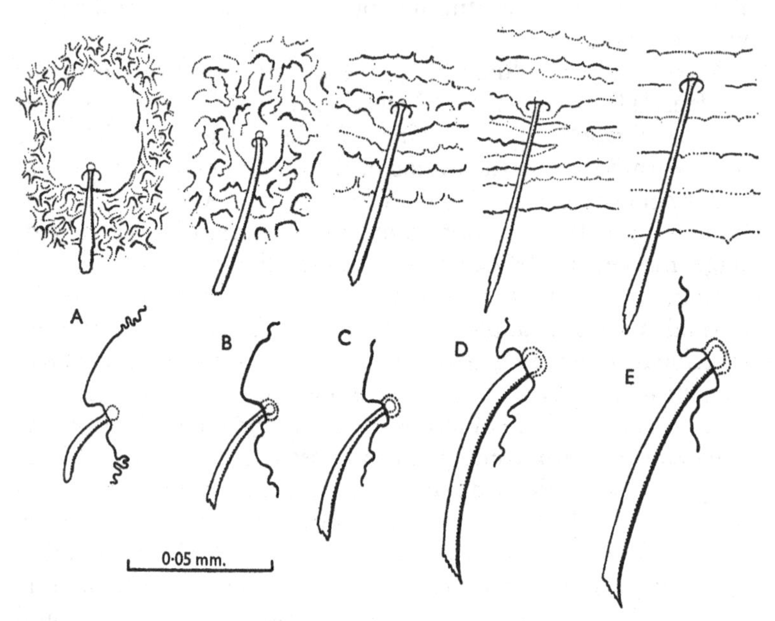

Fig. 39. Bristles intermediate between larval and adult type produced by decapitating *Rhodnius* 3rd-stage larvae during the critical period (cf. fig. 21). Upper row, bristles from dorsum of abdomen; lower row from margin of abdomen. A, normal 4th-stage larva; B, C, D, from forms showing increasing degrees of metamorphosis; E, normal adult for comparison. (Wigglesworth, 1934.)

through a series of growth stages, at only one of which are they capable of taking up from the circulating blood the precursor of the scale pigments. The timing of the growth stages is supposed to be different in different parts of the wing. If, therefore, the pigment precursors are liberated at different times they will be taken up only by those parts of the wing which have reached the right stage of growth at the corresponding time. The final wing pattern will represent a pattern of developmental rates.

The same general type of explanation is advanced by Goldschmidt (1938, 1940) to account for differentiation of the appendages. In the normal *Drosophila* the antenna-forming substance is supposed to become available precisely at the moment when the antennal anlage is ready to take up this substance and undergo differentiation to an antenna. But if the timing of these two processes is upset the antennal anlage may be in the receptive state at the time when the leg-forming substance becomes available and it therefore differentiates wholly or partially into a leg. Disturbances of this kind may result from the action of mutant genes: *aristopedia*, where the arista of the antenna is more or less leg-like; *proboscipedia*, where the labella of the proboscis are transformed into tarsus-like structures; *tetraptera*, where the halteres are changed to wings; or *tetraltera*, where wings are changed to halteres. Such genes are regarded by Goldschmidt as 'rate genes' which are influencing the velocity or the duration of some of the processes involved.

If these different growth processes were to be differently affected by temperature, then the action of genes producing hereditary homœosis might be expected to be influenced by the temperature of the environment. That proves to be the case. The 'penetrance' of *tetraptera* at 25° C. is 35 per cent; at 17° C. it is only 1 per cent (Astauroff, 1930). In *tetraltera* there is a converse effect: at 29° C. the penetrance is 1 per cent; at 14° C. 35 per cent. Food which prolongs development will likewise increase the penetrance of *tetraltera*. There are similar effects of temperature on *aristopedia*: cold treatment (14·5° C.) increases the expression of *aristopedia*, heat treatment (29° C.) decreases it (Villee, 1942); while the partial inhibition of growth in the antennal disc by treatment with colchicine deflects differentiation towards that of a normal arista (Vogt, 1947). In *proboscipedia* the modified oral lobes may come to resemble labrum, maxillary palps, antennae or tarsi, and yet changes in temperature may suppress the effect and lead to the formation of normal mouth-parts (Bridges and Dobzhansky, 1933). Cold treatment (15° C.) favours the appearance of arista-like appendages; heat treatment (29° C.) results in the development of tarsus-like appendages on the proboscis (Villee, 1944).

Not only is the manifestation of these abnormalities in the

mutant individuals influenced by temperature, but if a temperature shock, or some other adverse stimulus is administered at the appropriate or 'sensitive' period of growth these same abnormalities may appear in normal strains of *Drosophila*. If the *Drosophila* egg is exposed to a temperature shock during a period of 15 minutes at 3 hours after fertilization, a large proportion of the resulting flies have the metathorax more or less completely changed into a mesothorax, as brought about by the gene 'bithorax' (Maas, 1948). Exposure to ether vapour during this same period has the same effect (Gloor, 1947; Hadorn, 1948). Presumably this is the time (long before the imaginal discs first become visible at 16–18 hours) at which the corresponding gene exerts its action.

According to Goldschmidt (1935, 1945) it is only such morphological changes as result from differential effects on integrated reaction velocities which can be reproduced in this way as 'phenocopies' of particular mutants. But, as already pointed out in discussing prothetely (p. 90), although the Goldschmidt hypothesis often provides a very good formal explanation of the facts, it does not necessarily follow that the effects observed are brought about in this way. Throughout the whole process of scale formation in Lepidoptera, the developmental stages run parallel in the scales of all parts of the wing (Köhler and Feldotto, 1937), and if warm or cold tubes are applied to parts of the wing in pupae of *Vanessa*, they cause local delay or acceleration in development but they do not affect the ultimate colour pattern (Giersberg, 1929). So that if two competing processes are in fact at work in the determination of the wing pattern they must be processes occurring at an earlier stage than the visible differentiation of scale structure.

Similarly, the close study of differentiation in the antennae of *Drosophila* has provided no conclusive evidence in support of the Goldschmidt hypothesis. Throughout the last larval stage the anlage of the arista contains both arista- and leg-forming potencies. If a cold stimulus is applied to larvae of aristopedia mutants progressively later in the last stage, the arista character comes to predominate progressively, beginning at the distal extremity and proceeding proximally. So that when cold is applied in the latest stages only the most proximal part of the arista anlage is undetermined and capable of developing tarsus characteristics in response to cold (Vogt, 1946*b*).

It is during this same period that the labium in proboscipedia can be influenced by cold; but here, as we have seen, a cold stimulus has the opposite effect, it favours arista formation, whereas in aristopedia it favours tarsus formation (Vogt, 1946*c*). In the light of results of this kind it is unlikely that a specific evocator for tarsus or arista is being produced elsewhere in the body and diffusing into the anlage at some critical moment of development, as was originally suggested by Goldschmidt. As Villee (1944) points out, however, it is still possible to suppose that the evocators may be substances, produced within the nucleus of every cell of an imaginal disc, which can exert their determining influence only when the cell has reached the right state of competence.

But there is another possibility which is worth considering: that these morphological differences result from *quantitative* differences in a single evocator. That was the explanation put forward above for the determination of sensilla, at a high concentration, and dermal glands, at a low concentration of some substance produced in the epidermis of *Rhodnius*. It would be a surprising phenomenon if such striking morphological differences as those between a leg and an antenna could be brought about by a mere change in concentration of one component acting upon the substance of the undifferentiated cells. But we have seen that the characters of larva, pupa and imago in Lepidoptera, which are no less distinct, are induced by varying concentrations of the juvenile hormone.

As Waddington (1940*a*) has pointed out, the aristae in various grades of aristopedia are not simply intermediates between legs and aristae, but certain parts are strictly leg-like, while other parts are arista-like, the transition from one to the other occupying only a narrow zone. The tissues must choose between one type of development or the other. That does not preclude the possibility that the effects are brought about by different concentrations of a single evocator substance; for the same tendency towards 'all-or-none' development is seen not only in the differentiation of sensilla and dermal glands (p. 111), but also in the development of larval, pupal or imaginal characters (p. 94).

Where aristopedia has been induced in normal strains of *Drosophila* by exposing the larvae to the action of a nitrogen mustard for a brief period, the parts do in fact show all grades of transition between arista and tarsus (fig. 40) (Bodenstein and Abdel-Malek, 1949).

We have seen that in the action of the juvenile hormone, both the concentration and the timing of the secretion are important (p. 67). If a single modifier or evocator substance is concerned in the determination of the different types of appendage, the timing of the processes involved may be equally important.

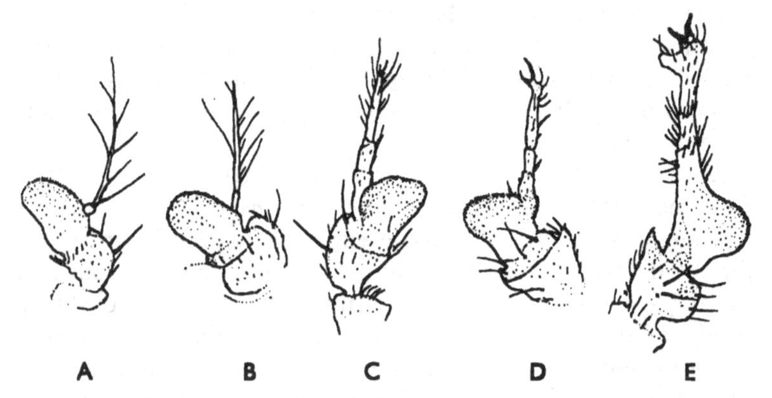

Fig. 40. A, normal antenna and arista of *Drosophila*. B–E, varying grades of aristopedia induced by brief treatment of the larva with a nitrogen mustard. (After Bodenstein and Abdel-Malek, 1949.)

And where these processes are differently affected by temperature, this may influence the determination of appendages, just as it influences the differentiation of larval and adult characters (p. 70) by disturbing the balance between the hormones controlling growth.

REGENERATION. According to the conception of differentiation developed in this chapter, an area in the undifferentiated substrate absorbs and unites with some substance, an inductor or modifier. The plasma thereby becomes determined for some particular type of development, and at the same time, by draining the inductor substance from the surrounding region, it inhibits a like determination in its vicinity. The zone of plasma determined in this way constitutes a 'field'. It is immaterial whether the modified plasma is divided into nucleated cells or whether it forms a non-nucleated continuum.

This change is illustrated in diagrammatic form in fig. 41. The zone '*A*' is the determined 'field'. This then proceeds to grow and within it the same type of change occurs, leading to a new and more specialized 'field' of determination '*B*'. And this in turn leads to the new field '*C*'. And so the process proceeds, with the uptake by an active centre of the materials necessary for a particular determination and the consequent

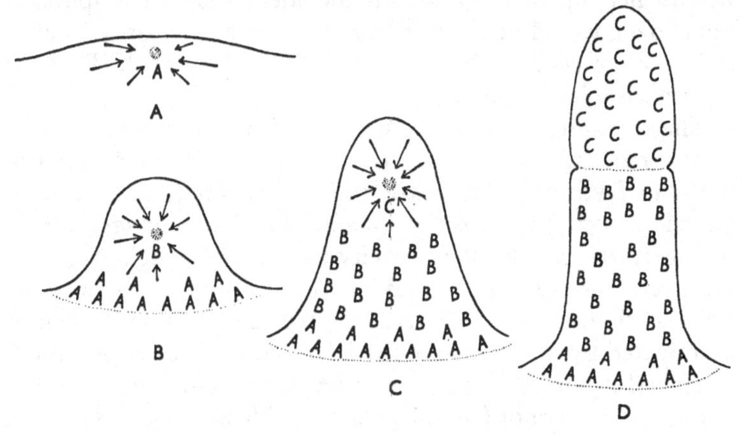

Fig. 41. Stages in the progressive determination in an appendage according to the hypothesis put forward. Explanation in the text.

inhibition in the surrounding zones of centres with the same potential activities. These potential centres are suppressed because they are deprived of their essential raw material. There is thus not only a limitation of potencies in the activated centre, there is also a positive evocation of faculties hitherto dormant. The process continues until terminal structures are produced, like the trichogen, tormogen or sense cell of the sensilla, in which, as the result of chemical determination followed by differentiating cell divisions the capacity of the cell is specialized for the performance of a single function.

This autonomous process of activation and suppression of chemical potentialities, leading to the production of a mosaic of 'fields', each with multiple potentialities and capable of differentiating further by the same procedure, may vary greatly in detail in different organs. It is characteristic of many insects that the general potentialities in the various fields of determination are not wholly lost as the successive steps of determination proceed within them. Suppose field '*A*' still persists at the base of the appendage in fig. 41 *d*; then if the distal parts of the appendage are removed it will be capable of repeating the process of determination and a new appendage will be regenerated—just as new dermal glands and plaques are regenerated during wound healing in *Rhodnius*. In the ideal case an amputated appendage would be expected to regenerate structures distal to the point of section, but not those lying proximal. That, in general, is what happens.

There is much variation in points of detail in regeneration as it occurs in different insects. In some Coleoptera (Hydrophilidae, Dytiscidae (Megušar, 1907), *Timarcha* (Abeloos, 1933; Bourdon, 1937)) there is no regeneration of appendages during the larval stages. But the capacity to form the imaginal appendage remains latent in the surrounding cells at the base of the limb; so that when the differentiation of the imaginal organism is activated by the change in the hormone balance at pupation, a normal adult appendage may be developed. On the other hand, in the larva of *Vanessa* a considerable zone of cells around the base of the complex spines must be removed if regeneration of the spine, by the larva, is to be prevented (Bodenstein, 1935).

The determination of the different levels in a limb is well illustrated by the experiments of Bodenstein (1935) in which the anterior appendages of the *Vanessa* larva were transplanted in place of partially removed posterior appendages (fig. 42). The combined material works together to form a harmonious limb with femur, tibia and tarsus; but the characters of the limb as a whole are intermediate: the resemblance to a fore-limb or a hind-limb depends on the relative quantities of the two elements. These results would suggest that determination of an appendage as a fore-leg or a hind-leg occurs very early in larval development; but the determination of the major components, femur, tibia and tarsus, is not complete until much later. It may be noted in passing that, just as in normal growth, the regenerated organ must reach a certain stage of development before it can undergo metamorphosis.

The parts of an appendage form an integrated whole, and one element in that integration is doubtless the inability of the limb at each level to develop the more proximal parts. If a limb which is capable of regeneration at any level is cut through and partially separated from the stump, or rotated through 180° so

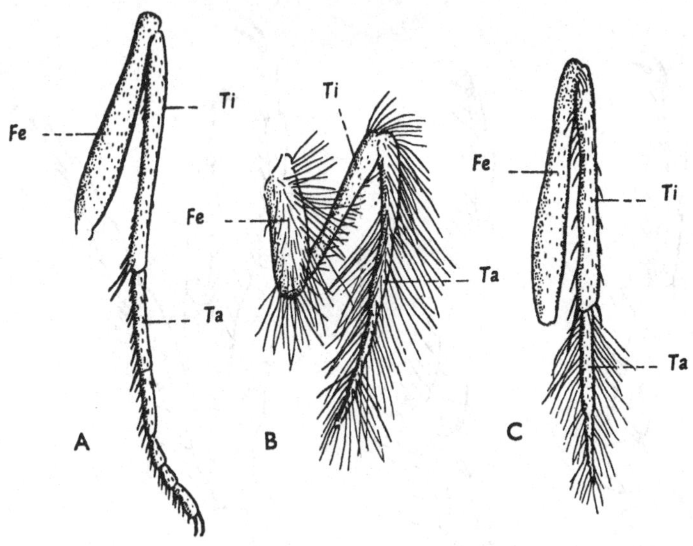

Fig. 42. A, normal hind-leg of *Vanessa urticae*; B, normal fore-leg; C, leg produced by implanting larval fore-leg on to the stump of the hind-leg. The tarsus combines scale characters of both fore and hind-legs. *Fe*, femur; *Ta*, tarsus; *Ti*, tibia. (After Bodenstein, 1935.)

that it does not join on properly to the stump (Bodenstein, 1937, 1941 *b*), a double regeneration occurs. At the distal end of the stump the distal parts of the appendage are regenerated as usual; and at the proximal end of the original distal fragment another distal piece is regenerated so that the final appendage has three branches.

If the regeneration of the partially detached fragment is considered (fig. 43), it can be seen that cells which grow out from the various faces of the limb will carry with them the characters for which these faces have been determined (just as the black spots or other types of cuticle spread during the repair of a wound (p. 52)). It follows, therefore, that the regenerate arising from this distal fragment will be a mirror image of it. That, in fact, is what always happens, in conformity with 'Bateson's law'.

In discussing homœosis the suggestion was put forward that the characters of an appendage may be controlled by the *amount*

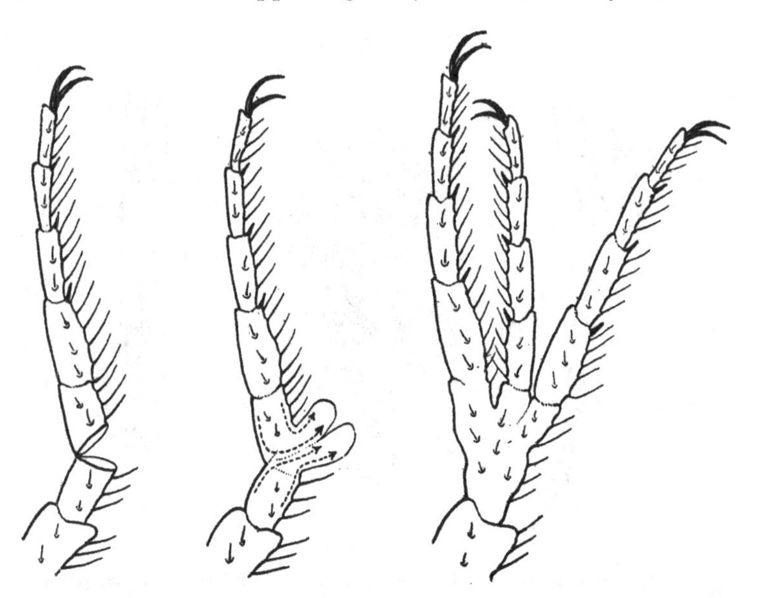

Fig. 43. Schema of regeneration in a partially detached appendage to illustrate the suggested explanation of Bateson's law. Explanation in text.

of the modifier or evocator available at the time of determination. If that were so it would not be surprising if homœosis were to occur during regeneration, since the amount of the evocator might be expected to be unduly small in the early stages of the process—just as, during the first moult after the repair of a burn in *Rhodnius*, dermal glands are developed but not sensilla (p. 109). That, in fact, is what happens; sometimes an extirpated antenna in *Sphodromantis* (Przibram, 1919) or *Dixippus* (Staudinger, 1930) will show the characters of a leg with a tarsus when it first regenerates and then continue its development as an antenna.

POLYMORPHISM. We have traced the process of determination and differentiation from the unicellular structures of the integument to the segmental appendages; and have shown that for purposes of description the same general conception of growth can be used throughout. That is not to say that the theory adopted is anything more than a rather schematic model of the process actually occurring in this highly complex medium. But a hypothesis which provides a consistent description of the facts, and can be used to predict the changes that occur is not without value.

It is no great step from the diverse form of appendages to the polymorphism of the individuals in a species. It is therefore worth considering how far these same ideas can be applied to the description of polymorphism in general. The example which has been studied in greatest detail in insects is the dimorphism of the sexes. Sex is determined primarily by the genetic constitution. That means that the genes are responsible for producing factors determining maleness and femaleness and that the balance of these sex determiners is such that if the genetic make-up is male, male characters are developed, and vice versa. The distribution of the genes responsible for these sex determiners is conceived as being somewhat different in *Drosophila* (Bridges, 1925) and *Lymantria* (Goldschmidt, 1931); but the general principle is the same.

The genes controlling sex in insects exert their action solely within the cells which carry them. If, therefore, the genetic constitution is different in different parts of the body sexual mosaics or 'gynandromorphs' are produced. But if the sex determiners are incorrectly balanced, as when different races of *Lymantria* (Goldschmidt, 1931), or of the psychid *Solenobia* (Seiler, 1937) are crossed, or in various species hybrids, 'intersexes' may result. In the most extreme forms it is possible to have normally functioning males or females with the genetic constitution of the opposite sex. In the less extreme cases the reversal of sex is partial and applies to some organs only.

In pure races of *Lymantria* with normally balanced determiners, it is possible to produce intersexual forms by exposure to extremes of temperature during development. These results are exactly comparable with the effects of environment on the penetrance of the genes of hereditary homœosis (p. 115) or on the production of 'phenocopies' (p. 116). In all these cases we have had occasion to note the 'all-or-none' effect in development, when the characters of sufficiently small units are considered: the cells develop either one character or the other. The same applies to sex: if the individual sensilla on the antennae of intersexual *Solenobia* are closely examined they are found to be either male or female, never intermediate (Seiler, 1951).

The same parallel is to be found in the effects of nutrition. Just as nutrition will influence the expression of *antennaless* or *aristopedia* in *Drosophila*, so the removal of nourishment from the tissues of some Hymenoptera and other insects (Salt, 1927, 1931; Koller, 1938; Hassan, 1939) by the presence of internal parasites (notably *Stylops*), will lead to the partial suppression of the genetic sex and the appearance of the opposite sexual characters. Presumably the defective nutrition of the cells, perhaps in respect to some specific substance, has changed the balance of the sex determiners within them.

There are other examples in which polymorphism is similarly influenced by nutrition. In the ant *Pheidole*, if larvae during a certain short period obtain such concentrated solid food that they can suddenly grow very rapidly, they develop into soldiers; if not, they become workers (Goetsch, 1937)—though here the effects of allometry consequent upon mere increase in size must be borne in mind (Wilson, 1953). Zander and Becker (1925) transferred worker larvae of the honey bee to queen cells at different stages of feeding. Those transferred up to 3½ days after hatching became normal queens. From this point onwards there was a very rapid change: within less than 12 hours the characters developed became those of workers. Intermediates were produced only in larvae which were changed over during this brief period; once the diet of pollen was started, no reversal was possible.

Such effects are still more striking in some endoparasitic insects in which the quantity and quality of food received is very exactly controlled by the size and nature of the host. The chalcid *Melittobia* develops in *Sceliphron* and other large hosts which will provide sustenance for 500–800 larvae. The first twenty or so to develop are brachypterous females and eyeless males. All the rest, receiving perhaps different food, develop into adults of normal form (Schmieder, 1939).

Males of *Trichogramma semblidis* are apterous when reared from the eggs of *Sialis* but winged and with striking differences in the morphology of the legs and antennae when reared from the eggs of *Ephestia* and other Lepidoptera (fig. 44) (Salt, 1937, 1938). Perhaps the dimorphism in the agamic and gamic generations

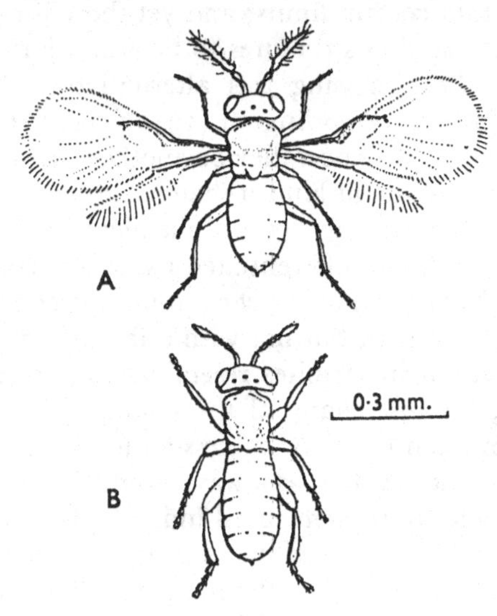

Fig. 44. Dimorphism in the male of *Trichogramma semblidis*. A, winged male reared in eggs of Lepidoptera; B, apterous male reared in eggs of *Sialis*. (After Salt, 1937.)

of female cynipids is due to their developing in different parts of the host plant (Salt, 1941).

In the ichneumonid *Gelis*, the female is always wingless, but the male is fully winged if it is fed adequately in the larval stage. On the other hand, if the male larvae are reared throughout in a small host, or are removed from a large host before they have finished feeding, they develop into micropterous males which differ from the normal not only in the development of the wings, but in the structure of the thorax, the size of the ocelli, the form of the endoskeleton, and the development, innervation and tracheal supply of the thoracic muscles. And there are no intermediates (Salt, 1952).

Other examples of the same kind could be adduced. The important point in the present context is that some nutritional change, which presumably means a change in the concentration of some single chemical substance, has led to an abrupt change in morphology in many different parts of the body. The whole phenomenon recalls the effects of the juvenile hormone in controlling metamorphosis. The level of juvenile hormone secretion must vary within certain limits; and yet there is a striking tendency for the resulting structures to be either larval, or pupal, or imaginal. In discussing this alternative or 'all-or-none' character of growth, Waddington (1940*b*) used the analogy of a stream running into one valley or another. In an attempt to find a parallel closer to the kind of change that may be happening in the tissues, the analogy with a chemical reaction was suggested (p. 110); the undifferentiated tissue was compared with a chemical substance which may combine with one or with two radicles of a given sort, but not with a fraction of a radicle.

Caste formation in termites resembles metamorphosis even more closely, because many of the castes, although typically terminal forms, can occur in succession to one another. Each morphological end result is, as always, controlled by genetic factors all of which are present in all individuals. The actual fate of the individual is controlled by a variety of factors in the complex environment which, for the most part, have not yet been analysed (Light, 1942). In *Calotermes*, for example (fig. 45), normal development leads through some six to ten larval stages to an indefinite caste termed the full-grown larva or pseudergate. These pseudergates can apparently make any number of 'stationary moults' without change in size. Alternatively, they may pass through two nymphal stages to the adult, through a 'pre-soldier' to a soldier, or directly to a neotenic sexual form. Soldiers and neotenics usually develop from pseudergates, but they can also arise from either of the two nymphal stages or from larvae of the 7th-stage onwards. Both nymphal stages can undergo regressive moults becoming pseudergates again and then develop in whatever direction the colony requires. This reversion prevents the appearance of winged adults in small colonies (Grassé and Noirot, 1947).

The determination of the direction of development in *Calotermes* is a 'group effect' in the sense used by Grassé (1946), but the factors concerned have been analysed only in the case of the neotenic sexual forms. If one sexual form of the pair is removed from a colony, any larvae at the appropriate or reactive stage become determined and developed into neotenics. The excess of these are eliminated by the other workers and a single one or pair is permitted to survive. If the original sexual pair in the colony is merely separated by a wire gauze barrier through which only the

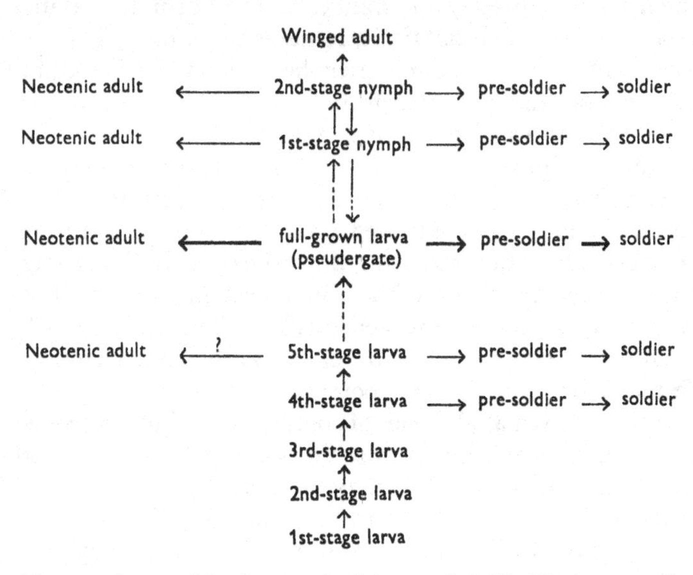

Fig. 45. Course of development in *Calotermes flavicollis*. The interrupted lines indicate the occurrence of several moults. (After Lüscher, 1952.)

antennae of the termites can be passed, neotenic individuals are produced—as though the pair had been removed completely. These results are interpreted as suggesting that some substance (often called a 'social hormone') is given off by the sexual pair and inhibits the determination of neotenics in the colony. This inhibition is no longer effective if a single gauze barrier is interposed; but all the neotenics produced are eliminated. If a pair is to be permitted to survive, a double barrier, which will prevent antennal contact, must be interposed (Lüscher, 1952*b*, 1953).

In the honey bee the queen produces a secretion which is urgently sought and licked from the surface of her body by the workers. The effect of this 'queen substance' is to inhibit the workers from building queen cells in the hive. It may also inhibit the ripening of the ovary and thus control the production of 'laying workers' (cf. p. 81) (Butler, 1954). Whether this material is to be regarded as a 'social hormone' which itself acts upon the metabolism of the worker, or as a sensory stimulus which leads indirectly to changes in the hormone balance of those workers which obtain it, remains to be proved.

CONCLUSION. In this chapter the subjects of differentiation, regeneration and polymorphism in insects have been treated very briefly and slightly. But enough has been written to show that metamorphosis is merely a special case of a general phenomenon, which is polymorphism in the widest sense. Polymorphism may be controlled by the genetic constitution of the individual. But often the differences in the body form of different individuals, like the differences in form in different parts of the body, arise in genetically uniform material. Such differences are controlled by chemical factors necessary for the realization of the latent potentialities.

Determination at all levels of complexity is thus controlled by the supply or deficiency of specific raw materials. The production of these raw materials may be upset by mutations in the nuclear genes; genetically controlled polymorphism is the result. Or, according to the view here brought forward, they may be produced locally within the tissues (and the nuclear genes are doubtless concerned in the process) and absorbed by one part in preference to another; such substances are termed inductors and they result in autonomous differentiation.

Or the raw material may be produced within some endocrine gland and circulate in the form of a hormone. The juvenile hormone or 'neotenin' from the corpus allatum is a substance of this kind the supply of which at high concentration is necessary to activate those cellular potentialities which provide the larval organism; at a much lower concentration the same hormone activates the precursor system of the pupal organism. The growth and moulting hormone of the thoracic gland is an essential raw material not only for the initiation of growth but for the activation of those cellular potentialities which provide for the differentiation of the adult organism. Finally, the raw material may be contained in the food. If the supply is below a certain threshold one form is produced, above that threshold another form.

The constancy of all these alternative or dichotomous paths of differentiation is one of the most striking features of every sort of polymorphism. It is an encouraging sign for those who are seeking a chemical explanation of growth and form.

REFERENCES

Abeloos, M. (1933). Sur la régenération des pattes chez la Coléoptère *Timarcha violaceo-nigra* de Geer. *C.R. Acad. Sci., Paris*, **113**, 17–19.

Abercrombie, W. F. (1936). Studies on cell number and the progression factor in the growth of Japanese beetle larvae (*Popillia japonica* Newman). *J. Morph.* **59**, 91–112.

Agrell, I. (1949). The variation in activity of apodehydrogenases during insect metamorphosis. *Acta physiol. scand.* **18**, 355–60.

Agrell, I. (1951). A contribution to the histolysis-histogenesis problem in insect metamorphosis. *Acta physiol. scand.* **23**, 179–86.

Allen, T. H. (1940). Cytochrome oxidase in relation to respiratory activity and growth of the grasshopper egg. *J. Cell. Comp. Physiol.* **16**, 149–63.

Alpatov, W. W. (1929). Growth and variation of the larvae of *Drosophila melanogaster*. *J. Exp. Zool.* **52**, 407–32.

Anglas, J. (1901). Quelques remarques sur les métamorphoses internes des Hyménoptères. *Bull. Soc. ent. Fr.*, pp. 104–7.

Aristotle. *Generation of Animals*, II, i; III, ix.

Arvy, L., Bounhiol, J.-J. and Gabe, M. (1953). Déroulement de la neurosécrétion protocérébrale chez *Bombyx mori* L. au cours du développement post-embryonnaire. *C.R. Acad. Sci., Paris*, **236**, 627–9.

Arvy, L. and Gabe, M. (1952). Données histophysiologiques sur les formations endocrines rétro-cérébrales de quelques Odonates. *Ann. Sci. nat. (Zool.)*, **14**, 345–74.

Arvy, L. and Gabe, M. (1953*a*). Données histophysiologiques sur la neurosécrétion chez quelques Ephéméroptères. *Cellule*, **55**, 203–22.

Arvy, L. and Gabe, M. (1953*b*). Données histophysiologiques sur la neurosécrétion chez les Paléoptères (Ephéméroptères et Odonates). *Z. Zellforsch.* **38**, 591–610.

Arvy, L. and Gabe, M. (1953*c*). Particularités histophysiologiques des glandes endocrines céphaliques chez *Tenebrio molitor* L. *C.R. Acad Sci., Paris*, **237**, 844–6.

Astauroff, B. L. (1930). Analyse der erblichen Störungsfälle der bilateralen Symmetrie im Zusammenhang mit der selbständigen Variabilität ähnlicher Strukturen. *Z. indukt. Abstamm.- u. VererbLehre*, **55**, 183–262.

Avery, O. T., MacLeod, C. M. and McCarty, M. (1944). Studies on the chemical nature of the substance inducing transformation of pneumococcal types. *J. Exp. Med.* **79**, 137–58.

Baitsell, G. A. (1940). A modern concept of the cell as a structural unit. *Amer. Nat.* **74**, 5–24.

Bataillon, E. (1893). La métamorphose du ver à soie et le déterminisme évolutif. *Bull. Sci. Fr. Belg.* **25**, 18–55.

References

BECK, S. D. (1950). Nutrition of the European corn borer *Pyrausta nubilalis* (Hbn.). ii. Some effects of diet on larval growth characteristics. *Physiol. Zoöl.* **23**, 353–61.

BECKER, E. (1941). Über Versuche zur Anreicherung und physiologischen Charakterisierung des Wirkstoffs der Puparisierung. *Biol. Zbl.* **61**, 360–88.

BECKER, E. and PLAGGE, (1939). Über das die Pupariumbildung auslösende Hormon der Fliegen. *Biol. Zbl.* **59**, 326–41.

BERGER, C. A. (1938). Multiplication and reduction of somatic chromosome groups as a regular developmental process in the mosquito, *Culex pipiens*. *Contr. Embryol. Carneg. Instn*, **27**, 209–34.

BERLESE, A. (1913). Intorno alle metamorphosi degli insetti. *Redia*, **9**, 121–36.

BODENSTEIN, D. (1933). Beintransplantationen an Lepidopteren Raupen. I. Transplantationen zur Analyse der Raupen- und Puppenhäutung. *Roux Arch. EntwMech. Organ.* **128**, 564–83.

BODENSTEIN, D. (1935). Beintransplantationen an Lepidopterenraupen. III. Zur Analyse der Entwicklungspotenzen der Schmetterlingsbeine. *Roux Arch. EntwMech. Organ.* **133**, 156–92.

BODENSTEIN, D. (1937). Beintransplantationen an Lepidopterenraupen. IV. Zur Analyse experimentell erzeugter Bein-Mehrfachbildungen. *Roux Arch. EntwMech. Organ.* **136**, 745–85.

BODENSTEIN, D. (1938). Untersuchungen zum Metamorphoseproblem. II. Entwicklungsrelationen in verschmolzenen Puppenteilen. *Roux Arch. EntwMech. Organ.* **137**, 636–60.

BODENSTEIN, D. (1939*a*). Investigations on the problem of metamorphosis. IV. Developmental relations of interspecific organ transplants in *Drosophila*. *J. Exp. Zool.* **82**, 1–30.

BODENSTEIN, D. (1939*b*). Investigations on the problem of metamorphosis. V. Some factors determining the facet number in the *Drosophila* mutant bar. *Genetics*, **24**, 494–508.

BODENSTEIN, D. (1941*a*). Investigations on the problem of metamorphosis. VII. Further studies on the determination of the facet number in *Drosophila*. *J. Exp. Zool.* **86**, 86–111.

BODENSTEIN, D. (1941*b*). Investigations on the problem of metamorphosis. VIII. Studies on leg determination in insects. *J. Exp. Zool.* **87**, 31–53.

BODENSTEIN, D. (1943*a*). Factors influencing growth and metamorphosis of the salivary gland in *Drosophila*. *Biol. Bull., Woods Hole*, **84**, 13–33.

BODENSTEIN, D. (1943*b*). Hormones and tissue competence in the development of *Drosophila*. *Biol. Bull. Woods Hole*, **84**, 34–58.

BODENSTEIN, D. (1946). A study of the relationship between organ and organic environment in the post embryonic development of the yellow fever mosquito. *Bull. Conn. Agric. Exp. Sta.* no. 501, pp. 100–14.

BODENSTEIN, D. (1947). Investigations on the reproductive system of *Drosophila*. *J. Exp. Zool.* **104**, 101–52.

BODENSTEIN, D. (1953*a*). Studies on the humoral mechanisms in growth and metamorphosis of the cockroach, *Periplaneta americana*. I. Transplantations of integumental structures and experimental parabioses. *J. Exp. Zool.* **123**, 189–232.

BODENSTEIN, D. (1953*b*). Studies on the humoral mechanisms in growth and metamorphosis of the cockroach *Periplaneta americana*. II. The function of the prothoracic gland and the corpus cardiacum. *J. Exp. Zool.* **123**, 413–33.

BODENSTEIN, D. (1953*c*). Studies on the humoral mechanisms in growth and metamorphosis of the cockroach, *Periplaneta americana*. III. Humoral effects on metabolism. *Biol. Bull., Woods Hole*, **124**, 105–15.

BODENSTEIN, D. (1953*d*). The role of hormones in moulting and metamorphosis. *Insect Physiology*, pp. 879–931. (K. D. Roeder ed.) New York, Wiley.

BODENSTEIN, D. and ABDEL-MALEK, A. (1949). The induction of aristopedia by nitrogen mustard in *Drosophila virilis*. *J. Exp. Zool.* **111**, 95–115.

BODENSTEIN, D. and SACKTOR, B. (1952). Cytochrome *c* oxidase activity during the metamorphosis of *Drosophila virilis*. *Science*, **116**, 299–300.

BODINE, J. H. (1934). The effect of cyanide on the oxygen consumption of normal and blocked embryonic cells (Orthoptera). *J. Cell. Comp. Physiol.* **4**, 397–404.

BODINE, J. H. (1941). The cell—some aspects of its functional ontogeny. *Amer. Nat.* **75**, 97–106.

BODINE, J. H. and BOELL, E. J. (1934). Respiratory mechanisms of normally developing and blocked embryonic cells (Orthoptera). *J. Cell. Comp. Physiol.* **5**, 97–113.

BODINE, J. H. and BOELL, E. J. (1936). The effect of ultracentrifuging on the respiratory activity of developing and blocked embryonic cells (Orthoptera). *J. Cell. Comp. Physiol.* **7**, 455–63.

BOUNHIOL, J. J. (1938). Recherches expérimentales sur le déterminisme de la métamorphose chez les Lépidoptères. *Bull. biol. Suppl.* **24**, 1–199.

BOUNHIOL, J. J. (1952*a*). L'achèvement de la métamorphose et la mue imaginale seraient commandés par la cerveau à la fin de la vie larvaire chez *Bombyx mori*. *C.R. Acad. Sci., Paris*, **235**, 671–2.

BOUNHIOL, J. J. (1952*b*). Nature probablement sécrétoire du facteur cérébral conditionnant la mue imaginale de *Bombyx mori* L. *C.R. Acad. Sci., Paris*, **235**, 747–8.

BOURDON, J. (1937). Sur la régénération des ébauches de quelques organes imaginaux chez le Coléoptère, *Timarcha goettingensis* L. *C. R. Soc. Biol., Paris*, **124**, 872–4.

BRADLEY, W. G. and ARBUTHNOT, K. D. (1938). The relation of host physiology to development of the Braconid parasite, *Chelonus annulipes* Wesmach. *Ann. Ent. Soc. Amer.* **31**, 359–65.

BRAUER, A. and TAYLOR, A. C. (1936). Experiments to determine the time and method of organization in Bruchid (Coleoptera) eggs. *J. Exp. Zool.* **73**, 127–52.

BRIDGES, C. B. (1925). Sex in relation to chromosomes and genes. *Amer. Nat.* **59**, 127–37.

References

BRIDGES, C. B. and DOBZHANSKY, T. (1933). The mutant 'proboscipedia' in *Drosophila melanogaster*—a case of hereditary homoösis. *Roux Arch. Entw-Mech. Organ.* **127**, 575–90

BÜCKMANN, D. (1953). Über den Verlauf und die Auslösung von Verhaltensänderungen und Umfärbungen erwachsener Schmetterlingsraupen. *Biol. Zbl.* **72**, 276–311.

BURTT, E. T. (1938). On the corpora allata of dipterous insects II. *Proc. Roy. Soc.* B, **126**, 210–23.

BUTENANDT, A. and KARLSON, P. (1954). Über die Isolierung eines Metamorphose-Hormons der Insekten in kristallisierter Form. *Z. Naturforsch.* **9** b, (in the press).

BUTENANDT, A., KARLSON, P. and HANNES, G. (1946). Über den 'Anti-bar-Stoff', einen genabhängigen, morphogenetischen Wirkstoff bei *Drosophila melanogaster*. *Biol. Zbl.* **65**, 41–51.

BUTLER, C.G. (1954). The method and importance of the recognition by a colony of honeybees (*A. mellifera*) of the presence of its queen. *Trans. R. Ent. Soc. Lond.* **105**, 11–29.

BUXTON, P. A. (1938). Studies on the growth of *Pediculus* (Anoplura). *Parasitology*, **30**, 65–84.

BYTINSKI-SALZ, H. (1933). Untersuchungen an Lepidopterenhybriden. II. Entwicklungsphysiologische Experimente über die Wirkung der disharmonischen Chromosomenkombinationen. *Roux Arch. EntwMech. Organ.* **129**, 356–78.

CAZAL, P. (1948). Les glandes endocrines rétro-cérébrales des insectes (étude morphologique). *Bull. biol. Suppl.* **32**, 1–227.

CHAUVIN, R. (1946). L'effect de groupe et la croissance larvaire des Blattes, du Grillon et du Phanéroptère. *Bull. Soc. Zool. Fr.* **71**, 39–48.

CHEFURKA, W. and WILLIAMS, C. M. (1952). Flavoproteins in relation to diapause and development in the Cecropia silkworm. *Anat. Rev.* **113**, no. 80.

CHEVAIS, S. (1943). Déterminisme de la taille de l'œil chez le mutant *Bar* de la Drosophile: intervention d'une substance diffusible spécifique. *Bull. biol.* **77**, 1–108.

CHRISTOPHERS, S. R. and CRAGG, F. W. (1922). On the so-called 'penis' of the bed-bug (*Cimex lectularius* L.) and on the homologies generally of the male and female genitalia of this insect. *Indian J. Med. Res.* **9**, 445–63.

COCKAYNE, E. A. (1941). Prothetely in a larva of *Smerinthus* hybr. *hybridus* Stephens (Lep.). *Proc. R. Ent. Soc. Lond.* A, **16**, 55–9.

COSTELLO, D. P. (1948). Ooplasmic segregation in relation to differentiation. *Ann. N.Y. Acad. Sci.* **49**, 663–83.

COUSIN, G. (1935). Sur les phénomènes de néoténie chez *Acheta campestris* L. et ses hybrides. *C.R. Acad. Sci., Paris*, **200**, 970–2.

CRAMPTON, H. E. (1899). An experimental study upon Lepidoptera. *Arch. EntwMech. Org.* **9**, 293–318.

DAY, M. F. (1943). The function of the corpus allatum in muscoid Diptera. *Biol. Bull., Woods Hole*, **84**, 127–40.

DEROUX-STRALLA, D. (1948). Recherches expérimentales sur le rôle des 'glandes ventrales' dans la mue et la métamorphose, chez *Aeschna cyanea* Müll. (Odonata). *C.R. Acad. Sci., Paris*, **227**, 1277–8.

DEWITZ, J. (1905). Untersuchungen über die Verwandlung der Insektenlarven. II. *Arch. Anat. Physiol., Lpz., Abt. Physiol.*, pp. 389–415.

DEWITZ, J. (1916). Bedeutung der oxydierenden Fermente (Tyrosinase) für die Verwandlung der Insektenlarven. *Zool. Anz.* **47**, 123–4.

DU BOIS, A. M. (1938). La détermination de l'ébauche embryonnaire chez *Sialis lutaria* L. (Megaloptera). *Rev. suisse Zool.* **45**, 1–92.

DU BOIS, A. M. and GEIGY, R. (1935). Beiträge zur Oekologie, Fortpflanzungsbiologie und Metamorphose von *Sialis lutaria* L. *Rev. suisse Zool.* **42**, 169–248.

DUPONT-RAABE, M. (1951). Étude morphologique et cytologique du cerveau de quelques phasmides. *Bull. Soc. zool. Fr.* **76**, 386–97.

DUPONT-RAABE, M. (1952). Contribution à l'étude du rôle endocrine du cerveau et notamment de la *pars intercerebralis* chez les Phasmides. *Arch. Zool. exp. gen.* **89**, 128–38.

DYAR, H. G. (1890). The number of molts in Lepidopterous larvae. *Psyche*, **5**, 420–2.

EDLBACHER, S. (1946). Das Ganzheitsproblem in der Biochemie. *Experientia* **2**, 7–18.

ENZMANN, E. V. and HASKINS, C. P. (1938). The development of the imaginal eye in the larva of *Drosophila melanogaster*. *J. Morph.* **63**, 63–72.

EPHRUSSI, B. (1942). Chemistry of 'eye colour hormones' of *Drosophila*. *Quart. Rev. Biol.* **17**, 326–38.

EPHRUSSI, B. (1953). *Nucleo-cytoplasmic Relations in Micro-organisms; Their Bearing on Cell Heredity and Differentiation.* Oxford: University Press.

EVANS, A. C. (1936). Histolysis of muscle in the pupa of the blow-fly *Lucilia sericata* Meig. *Proc. R. Ent. Soc. Lond.* A, **11**, 52–4.

EWEST, A. (1937). Struktur und erste Differenzierung im Ei des Mehlkäfers *Tenebrio molitor*. *Roux Arch. EntwMech. Organ.* **135**, 689–752.

FAURÉ-FREMIET, E. (1948). Les mécanismes de la morphogénèse chez les ciliés. *Folia biotheor., Leiden*, no. III, 25–58.

FRAENKEL, G. (1935*a*). A hormone causing pupation in the blowfly *Calliphora erythrocephala*. *Proc. Roy. Soc.* B, **118**, 1–12.

FRAENKEL, G. (1935*b*). Observations and experiments on the blow-fly (*Calliphora erythrocephala*) during the first day after emergence. *Proc. Zool. Soc. Lond.* pp. 893–904.

FUKUDA, S. (1939). Acceleration of development of silkworm ovary by transplantation into young pupa. *Proc. Imp. Acad. Japan*, **15**, 19–21.

FUKUDA, S. (1940*a*). Induction of pupation in silkworm by transplanting the prothoracic gland. *Proc. Imp. Acad. Japan*, **16**, 414–16.

FUKUDA, S. (1940*b*). Hormonal control of moulting and pupation in the silkworm. *Proc. Imp. Acad. Japan*, **16**, 417–20.

FUKUDA, S. (1944). The hormonal mechanism of larval molting and metamorphosis in the silkworm. *J. Fac. Sci. Tokyo Univ.* sec. IV, **6**, 477–532.

References

FUKUDA, S. (1951). The production of the diapause eggs by transplanting the suboesophageal ganglion in the silkworm. *Proc. Imp. Acad. Japan*, **27**, 672–7.

FUKUDA, S. (1952). Function of the pupal brain and suboesophageal ganglion in the production of non-diapause and diapause eggs in the silkworm. *Ann. Zool. Jap.* **25**, 149–55.

FURUKAWA, H. (1935). Can the skin of imago be made to moult? *Proc. Imp. Acad. Japan*, **11**, 158–60.

GABE, M. (1953*a*). Quelques acquisitions récentes sur les glandes endocrines des Arthropodes. *Experientia*, **9**, 352–6.

GABE, M. (1953*b*). Données histologiques sur les glandes endocrines céphaliques de quelques Thysanoures. *Bull. Soc. zool. Fr.* **78**, 177.

GAINES, J. C. and CAMPBELL, F. L. (1935). Dyar's rule as related to the number of instars of the corn ear worm, *Heliothis obsoleta* (Fab.), collected in the field. *Ann. Ent. Soc. Amer.* **28**, 445–61.

GEIGY, R. (1931). Erzeugung rein imaginaler Defekte durch ultraviolette Eibestrahlung bei *Drosophila melanogaster. Roux Arch. EntwMech. Organ.* **125**, 406–47.

GEIGY, R. (1941). Die Metamorphose als Folge gewebsspezifischer Determination. *Rev. suisse Zool.* **48**, 483–94.

GIERSBERG, H. (1929). Die Färbung der Schmetterlinge. I. *Z. vergl. Physiol.* **9**, 523–52.

GILLETT, J. D. (1935). The genital sterna of the immature stages of *Rhodnius prolixus* (Hemiptera). *Trans. R. Ent. Soc. Lond.* **83**, 1–5.

GLOOR, H. (1947). Phänokopie-Versuche mit Äther an *Drosophila. Rev. suisse Zool.* **54**, 637–712.

GOETSCH, W. (1937). Die Entstehung der 'Soldaten' im Ameisenstaat. *Naturwissenschaften*, **25**, 803–8.

GOLDSCHMIDT, R. (1923). Einige Materialien zur Theorie der abgestimmten Reaktionsgeschwindigkeiten. *Arch. mikr. Anat.* **98**, 292–313.

GOLDSCHMIDT, R. (1927). *Physiologische Theorie der Vererbung*. Berlin: Springer.

GOLDSCHMIDT, R. (1931). Analysis of intersexuality in the gipsy-moth. *Quart. Rev. Biol.* **6**, 125–142.

GOLDSCHMIDT, R. (1935). Gen und Aussencharakter. III. *Biol. Zbl.* **55**, 535–554.

GOLDSCHMIDT, R. (1938). *Physiological Genetics*. New York: McGraw-Hill.

GOLDSCHMIDT, R. (1940). *The Material Basis of Evolution*. Boston: Yale University Press.

GOLDSCHMIDT, R. (1945). Additional data on phenocopies and genic action. *J. Exp. Zool.* **100**, 193–201.

GOMORI, G. (1941). Observations with differential stains on human islets of Langerhans. *Amer. J. Path.* **17**, 395–406.

GORDON, C. and SANG, J. H. (1941). The relation between nutrition and exhibition of the gene Antennaless (*Drosophila melanogaster*). *Proc. Roy. Soc.* B, **130**, 151–84.

GRASSÉ, P. P. (1946). La structure des sociétés d'Invertébrés. *Rev. suisse Zool.* **53**, 432–41.

GRASSÉ, P. P. and NOIROT, C. (1947). Le polymorphisme social du termite à cou jaune (*Calotermes flavicollis* F.). Les faux-ouvriers ou *pseudergates* et les mues régressives. *C.R. Acad. Sci., Paris*, **224**, 219–21.

GUNN, D. L. and KNIGHT, R. H. (1945). The biology and behaviour of *Ptinus tectus* Boie (Coleoptera, Ptinidae), a pest of stored products. VI. Culture conditions. *J. Exp. Biol.* **21**, 132–43.

HACHLOW, V. (1931). Zur Entwicklungsmechanik der Schmetterlinge. *Roux. Arch. EntwMech. Organ.* **125**, 26–49.

HADORN, E. (1948). Genetische und entwicklungsphysiologische Probleme der Insektenontogenese. *Folia biotheor., Leiden*, no. 3, pp. 109–26.

HADORN, E., BERTANI, G. and GALLERA, J. (1949). Regulationsfähigkeit und Feldorganisation der männlichen Genital-Imaginalscheibe von *Drosophila melanogaster. Roux Arch. EntwMech. Organ.* **144**, 31–70.

HADORN, E. and NEEL, J. (1938). Der hormonale Einfluss der Ringdrüse (corpus allatum) auf die Pupariumbildung bei Fliegen. *Roux Arch. EntwMech. Organ.* **138**, 281–304.

HAECKEL, E. (1857). Ueber die Gewebe des Flusskrebses. *Arch. Anat. Physiol., Lpz.*, pp. 469–568.

HAHN, J. (1932). Životní prostor a jeho vliv na vývoj *Tenebrio molitor. Acta Soc. Ent. Bohem. (Čsl.)*, pp. 176–88.

HANDLIRSCH, A. (1927). Die postembryonale Entwicklung. Schröder's *Handbuch der Entomologie*, **1**, 1117–85.

HANSTRÖM, B. (1938). Zwei Probleme betreffs der hormonalen Lokalisation im Insektenkopf. *Acta Univ. lund.*, N.F., Avd. 2, **39**, 1–17.

HANSTRÖM, B. (1940). Inkretorische Organe, Sinnesorgane und Nervensystem des Kopfes einiger niederer Insektenordnungen. *K. svenska VetensAkad. Handl.* **18**, 3–265.

HANSTRÖM, B. (1941). Einige Parallelen im Bau und in der Herkunft der inkretorischen Organe der Arthropoden und der Vertebraten. *Acta Univ. lund.*, N.F., Avd. 2, **37**, 1–19.

HANSTRÖM, B. (1949). Three principal incretory organs in the animal kingdom. *Bull. biol.* Suppl. **33**, 182–209.

HANSTRÖM, B. (1953). Neurosecretory pathways in the head of crustaceans, insects and vertebrates. *Nature, Lond.*, **171**, 72–3.

HARVEY, W. (1651). *Generatione Animalium.* exerc. XLV and L. Amsterdam.

HASSAN, A. I. (1939). The biology of some British Delphacidae (Homopt.) and their parasites with special reference to the Strepsiptera. *Trans. R. Ent. Soc. Lond.* **89**, 345–84.

HENKE, K. (1933). Zur Morphologie und Entwicklungsphysiologie der Tierzeichnungen. *Naturwissenschaften*, **21**, 633–9, 654–9, 665–73, 683–90.

HENSON, H. (1946). The theoretical aspect of insect metamorphosis. *Biol. Rev.* **21**, 1–14.

HINTON, H. E. (1948). On the origin and function of the pupal stage. *Trans. R. Ent. Soc. Lond.* **99**, 395–409.

HOWLAND, R. B. and CHILD, G. P. (1935). Experimental studies on development in *Drosophila melanogaster*. 1. Removal of protoplasmic materials during late cleavage and early embryonic stages. *J. Exp. Zool.* **70**, 415–24.

References

Howland, R. B. and Sonnenblick, B. P. (1936). Experimental studies on development in *Drosophila melanogaster*. 2. Regulation in the early egg. *J. Exp. Zool.* **73**, 109–52.

Huxley, J. S. (1932). *Problems of Relative Growth*. London: Methuen.

Imms, A. D. (1925). *Textbook of Entomology*. London: Methuen.

Imms, A. D. (1937). *Recent Advances in Entomology*. London: Churchill.

Ito, H. (1918). On the glandular nature of the corpora allata of the Lepidoptera. *Bull. Imp. Tokyo Sericultural Coll.* **1**, 63–103.

Jensen, P. B. (1948). A determination theory. *Physiol. Plant.* **1**, 156–69.

Joly, P. (1945*a*). La fonction ovarienne et son contrôle humoral chez les Dytiscidés. *Arch. Zool. exp. gén.* **84**, 49–164.

Joly, P. (1945*b*). Les corrélations humorales chez les insectes. *Année biol.* **21**, 1–34.

Jones, B. M. (1953). Activity of the incretory centres of *Locustana pardalina* during embryogenesis: Function of the prothoracic glands. *Nature, Lond.*, **172**, 551.

Kaiser, P. (1949). Histologische Untersuchungen über die Corpora allata und Prothoraxdrüsen der Lepidopteren in Bezug auf ihre Funktion. *Roux Arch. EntwMech. Organ.* **144**, 99–131.

Karlson, P. (1954). Biochemische Probleme der Insektenmetamorphose. *Verh. Deutsch. Zool. Ges.* (in the press).

Karlson, P. and Hanser, G. (1952). Über die Wirkung des Puparisierungshormons bei der Wildform und der Mutante *lgl* von *Drosophila melanogaster*. *Z. Naturf.* **7**b, 80–3.

Karlson, P. and Hanser, G. (1953). Bildungsort und Erfolgsorgan des Puparisierungshormons der Fliegen. *Z. Naturf.* **8**b, 91–6.

Ketchel, M. and Williams, C. M. (1952). Relationship between the hemagglutination factor and the growth and differentiation hormone in the *Cecropia* silkworm. *Anat. Rec.* **113**, no. 83.

Key, K. H. L. (1936). Observations on rate of growth, coloration, and the abnormal six-instar life cycle in *Locusta migratoria migratorioides* R. and F. *Bull. Ent. Res.* **27**, 77–85.

Key, K. H. L. and Edney, E. B. (1936). Precocious adults resulting from the omission of the fifth instar in *Locusta migratoria migratorioides* R. and F. *Proc. R. Ent. Soc. Lond.* **11**, 55–8.

Klatt, B. (1919). Keimdrüsentransplantationen beim Schwammspinner. *Z. indukt. Abstam.- und VererbLehre*, **22**, 1–50.

Klein, H. Z. (1932). Studien zur Oekologie und Epidemiologie der Kohlweissling. 1. Der Einfluss der Temperatur und Luftfeuchtigkeit auf Entwicklung und Mortalität von *Pieris brassicae* L. *Z. angew. Ent.* **19**, 395–448.

Kogure, M. (1933). The influence of light and temperature on certain characters of the silkworm *Bombyx mori*. *J. Dep. Agric. Kyushu Univ.* **4**, 1–93.

Köhler, W. (1932). Die Entwicklung der Flügel bei der Mehlmotte *Ephestia kühniella* Zeller, mit besonderer Berücksichtigung des Zeichnungsmusters. *Z. Morph. Ökol. Tiere*, **24**, 582–681.

Köhler, W. and Feldotto, W. (1937). Morphologische und experimentelle Untersuchungen über Farbe, Form und Struktur der Schuppen von *Vanessa urticae* und ihre gegenseitigen Beziehungen. *Roux Arch. EntwMech. Organ.* **136**, 313–99.

Koller, G. (1938). *Hormone bei wirbellosen Tieren.* Leipzig.

Kopeč, S. (1911). Untersuchungen über Kastration und Transplantation bei Schmetterlingen. *Arch. EntwMech. Org.* **33**, 1–116.

Kopeč, S. (1917). Experiments on metamorphosis of insects. *Bull. int. Acad. Cracovie* (B), pp. 57–60.

Kopeč, S. (1922). Studies on the necessity of the brain for the inception of insect metamorphosis. *Biol. Bull., Woods Hole*, **42**, 322–42.

Kopeč, S. (1923). The influence of the nervous system on the development and regeneration of muscles and integument in insects. *J. Exp. Zool.* **37**, 15–25.

Kopeč, S. (1924). Studies on the influence of inanition on the development and duration of life in insects. *Biol. Bull., Woods Hole*, **46**, 1–21.

Kopeč, S. (1927). Über die Entwicklung der Insekten unter dem Einfluss der Vitaminzugabe. *Biol. gen.* **3**, 375–84.

Kornhauser, S. I. (1919). The sexual characteristics of the membracid, *Thelia bimaculata* (Fabr.). I. External changes induced by *Aphelopus theliae* (Gahan). *J. Morph.* **32**, 531–636.

Kowalewsky, A. (1887). Beiträge zur Kenntnis der nachembryonalen Entwicklung der Musciden. *Z. wiss. Zool.* **45**, 542–94.

Krause, G. (1953). Die Aktionsfolge zur Gestaltung des Keimstreifs von *Tachycines* (Saltatoria), insbesondere das morphogenetische Konstruktionsbild bei Duplicitas parallela. *Roux Arch. EntwMech. Organ.* **146**, 275–370.

Kreyenberg, J. (1929). Experimentell-biologische Untersuchungen über *Dermestes lardarius* L. und *D. vulpinus* F. Ein Beitrag zur Frage nach der Inconstanz der Häutungszahlen bei Coleopteren. *Z. angew. Ent.* **14**, 140–88.

Krumiņš, R. (1952). Die Borstenentwicklung bei der Wachsmotte *Galleria mellonella* L. *Biol. Zbl.* **71**, 183–210.

Kühn, A. and Piepho, H. (1936). Über hormonale Wirkungen bei der Verpuppung der Schmetterlinge. *Ges. Wiss. Göttingen, Nachrichten a. d. Biol.* **2**, 141–54.

Kühn, A. and Piepho, H. (1938). Die Reaktionen der Hypodermis und der Versonschen Drüsen auf das Verpuppungshormon bei *Ephestia kühniella. Biol. Zbl.* **58**, 12–51.

Kühn, A. and Piepho, H. (1940). Über die Ausbildung der Schuppen in Hauttransplantaten von Schmetterlingen. *Biol. Zbl.* **60**, 1–22.

Lamborn, W. A. (1914). The retention of spaces for the 'tails' in the pupae of the tailless females of *Papilio dardanus. Proc. Ent. Soc. Lond.* p. lxvii.

Lee, H. Tsui-Ying (1948). A comparative morphological study of the prothoracic glandular bands of some Lepidopterous larvae with special reference to their innervation. *Ann. Ent. Soc. Amer.* **41**, 200–5.

References

LEES, A. D. (1954). *The Physiology of Diapause in Insects*. Cambridge University Press.

LEES, A. D. and PICKEN, L. E. R. (1945). Shape in relation to fine structure in the bristles of *Drosophila melanogaster*. *Proc. Roy. Soc.* B, **132**, 396–423.

v. LENGERKEN, H. (1932). Nachhinkende Entwicklung und ihre Folgeerscheinungen beim Mehlkäfer. *Jena. Z. Naturw.* **67**, 260–73.

DE LERMA, B. (1942). Ricerche sperimentali sulle metamorfosi dei Ditteri. *Boll. Zool.* **13**, no. 3.

DE LERMA, B. (1950). Endocrinologia degli insetti. *Boll. Zool.* **17** (suppl.), 68–192.

DE LERMA, B. (1951). Note originali e critiche sulla morphologia comparata degli organi frontali degli Artropodi. *Annu. Ist. Zool. Univ. Napoli*, **3**, 1–23.

L'HELIAS, C. (1953*a*). Rôle des corpora allata dans le métabolisme des glucides, de l'azote et des lipides chez le phasme *Dixippus morosus*. *C.R. Acad. Sci., Paris*, **236**, 2164–6.

L'HELIAS, C. (1953*b*). Étude comparée de l'azote total et de l'azote non proteinique chez le phasme *Dixippus morosus* après ablation des corpora allata. *C.R. Acad. Sci., Paris*, **236**, 2489–91.

LHOSTE, J. (1951). Les modifications de structure des glandes ventrales chez *Forficula auricularia* L. au cours du développement. *Bull. Soc. zool. Fr.* **75**, 285–92.

LIGHT, S. F. (1942). The determination of the castes of social insects. *Quart. Rev. Biol.* **17**, 312–26; **18**, 46–63.

LILLIE, F. R. (1909). Polarity and bilaterality of the annelid egg. *Biol. Bull., Woods Hole*, **16**, 54–79.

LILLIE, F. R. (1929). Embryonic segregation and its role in the life history. *Roux Arch. EntwMech. Organ.* **118**, 499–533.

LONG, D. B. (1953). Effects of population density on larvae of Lepidoptera. *Trans. R. Ent. Soc. Lond.* **104**, 543–85.

LUBBOCK, J. (1883). *The Origin and Metamorphosis of Insects*. London: Macmillan.

LUDWIG, D. (1953). Cytochrome oxidase activity during diapause and metamorphosis of the Japanese beetle (*Popillia japonica* Newman). *J. Gen. Physiol.* **36**, 751–7.

LÜSCHER, M. (1944). Experimentelle Untersuchungen über die larvale und die imaginale Determination im Ei der Kleidermotte (*Tineola bisselliella* Hum.). *Rev. suisse Zool.* **51**, 531–627.

LÜSCHER, M. (1951). Die Produktion und Elimination von Ersatzgeschlechtstieren bei der Termite *Kalotermes flavicollis* Fabr. *Z. vergl. Physiol.* **34**, 123–41.

LÜSCHER, M. (1952*a*). Die Ursachen der tierischen Regeneration. *Experientia*, **8**, 81.

LÜSCHER, M. (1952*b*). Untersuchungen über das individuelle Wachstum bei der Termite *Kalotermes flavicollis* Fabr. (Ein Beitrag zum Kastenbildungsproblem.) *Biol. Zbl.* **71**, 529–43.

LÜSCHER, M. (1953). Kann die Determination durch eine monomolekulare Reaktion ausgelöst werden? *Rev. suisse Zool.* **60**, 524–8.

LYONET, P. (1762). *Traité anatomique de la chenille qui ronge le bois de saule.* La Haye.

MAAS, A. H. (1948). Über die Auslösbarkeit von Temperaturmodifikationen während der Embryonal-Entwicklung von *Drosophila melanogaster* Meigen. *Roux Arch. EntwMech. Organ.* **143**, 515–72.

MACDONALD, S. and BROWN, A. W. A. (1952). Cytochrome oxidase and cyanide sensitivity of the larch sawfly during metamorphosis. *Ann. Rep. Ent. Soc. Ont.* **83**, 30–4.

MAGNUSSEN, K. (1933). Untersuchungen zur Entwicklungsphysiologie des Schmetterlingsflügels. *Roux Arch. EntwMech. Organ.* **128**, 447–97.

MEGUŠAR, F. (1907). Die Regeneration der Coleopteren. *Arch. EntwMech. Org.* **25**, 148–234.

MELLANBY, K. (1938). Diapause and metamorphosis of the blowfly, *Lucilia sericata* Meig. *Parasitology,* **30**, 392–402.

MENDES, M. V. (1948). Histology of the corpora allata of *Melanoplus differentialis* (Orthoptera: Saltatoria). *Biol. Bull., Woods Hole,* **94**, 194–207.

MEYER, J. H. (1953). Die Bluttransfusion als Mittel zur Überwindung letaler Keimkombination bei Lepidopteren-Bastarden. *Z. wien. ent. Ges.* **38**, 41–80.

MÜLLER, F. (1869). *Facts and Arguments for Darwin.* London.

MURRAY, F. V. and TIEGS, O. W. (1935). The metamorphosis of *Calandra oryzae. Quart. J. Micr. Sci.* **77**, 405–95.

MÜSSBICHLER, A. (1952). Die Bedeutung äusserer Einflüsse und der Corpora allata bei der Afterweiselentstehung von *Apis mellifica. Z. vergl. Physiol.* **34**, 207–21.

NABERT, A. (1913). Die *Corpora allata* der Insekten. *Z. wiss. Zool.* **104**, 181–358.

NAGEL, R. H. (1934). Metathetely in larvae of the confused flour beetle (*Tribolium confusum* Duval). *Ann. Ent. Soc. Amer.* **27**, 425–8.

NAYAR, K. K. (1953). Neurosecretion in *Iphita limbata* Stal. *Curr. Sci.* **22**, 149.

NAYAR, K. K. (1954). Metamorphosis in the integument of caterpillars with omission in the pupal stage. *Proc. R. Ent. Soc. Lond.* A. (In the press.)

NOVÁK, V. J. A. (1951*a*). New aspects of the metamorphosis of insects. *Nature, Lond.,* **167**, 132–3.

NOVÁK, V. J. A. (1951*b*). The metamorphosis hormones and morphogenesis in *Oncopeltus fasciatus* Dal. *Mém. Soc. zool. tchécosl.* **15**, 1–47.

NÜESCH, H. (1952). Über den Einfluss der Nerven auf die Muskelentwicklung bei *Telea polyphemus* (Lep.). *Rev. suisse Zool.* **59**, 294–301.

OERTEL, E. (1930). Metamorphosis in the honeybee. *J. Morph.* **50**, 295–340.

OGURA, S. (1933). Erblichkeitsstudien am Seidenspinner *Bombyx mori* L. III. Genetische Untersuchung der Häutung. *Z. indukt. Abstamm.- u. VererbLehre,* **64**, 205–68.

PAPPENHEIMER, A. M. and WILLIAMS, C. M. (1952). The effects of the diphtheria toxin on the Cecropia silkworm. *J. Gen. Physiol.* **35**, 727–40.

PASSONNEAU, J. V. and WILLIAMS, C. M. (1953). The moulting fluid of the Cecropia silkworm. *J. Exp. Biol.* **30**, 545–60.

PÉREZ, C. (1902). Contribution à l'étude des métamorphoses. *Bull. sci. Fr. Belg.* **37**, 195–427.

PÉREZ, C. (1910). Recherches histologiques sur la métamorphose des Muscides (*Calliphora erythrocephala* Mg.). *Arch. Zool. exp. gén.* **4**, 1–274.

PFEIFFER, I. W. (1939). Experimental study of the function of the corpora allata in the grasshopper, *Melanoplus differentialis*. *J. Exp. Zool.* **82**, 439–61.

PFEIFFER, I. W. (1945*a*). Effect of the corpora allata on the metabolism of adult female grasshoppers. *J. Exp. Zool.* **99**, 183–233.

PFEIFFER, I. W. (1945*b*). The influence of the corpora allata over the development of nymphal characters in the grasshopper *Melanoplus differentialis*. *Trans. Conn. Acad. Arts Sci.* **36**, 489–515.

PFLUGFELDER, O. (1937). Bau, Entwicklung und Funktion der Corpora allata und cardiaca von *Dixippus morosus* Br. *Z. wiss. Zool.* A, **149**, 477–512.

PFLUGFELDER, O. (1938*a*). Untersuchungen über histologischen Veränderungen und das Kernwachstum der Corpora allata von Termiten. *Z. wiss. Zool.* **150**, 451–67.

PFLUGFELDER, O. (1938*b*). Weitere experimentelle Untersuchungen über die Funktion der Corpora allata von *Dixippus morosus* Br. *Z. wiss. Zool.* **151**, 149–91.

PFLUGFELDER, O. (1939*a*). Wechselwirkungen von Drüsen innerer Sekretion bei *Dixippus morosus* Br. *Z. wiss. Zool.* **152**, 384–408.

PFLUGFELDER, O. (1939*b*). Beeinflüssung von Regenerationsvorgängen bei *Dixippus morosus* Br. durch Exstirpation und Transplantation der Corpora allata. *Z. wiss. Zool.* **152**, 159–84.

PFLUGFELDER, O. (1940). Austausch verschieden alter C. allata bei *Dixippus morosus* Br. *Z. wiss. Zool.* **153**.

PFLUGFELDER, O. (1947*a*). Geschwulstartige Wucherungen embryonaler Transplantate in *Carausius* (*Dixippus*) *morosus* nach experimenteller Störung des Hormonhaushaltes. *Biol. Zbl.* **66**, 170–78.

PFLUGFELDER, O. (1947*b*). Über die Ventraldrüsen und einige andere inkretorsiche Organe des Insektenkopfes. *Biol. Zbl.* **66**, 211–35.

PFLUGFELDER, O. (1948). Volumetrische Untersuchungen an den Corpora allata der Honigbiene, *Apis mellifica* L. *Biol. Zbl.* **67**, 223–41.

PFLUGFELDER, O. (1949). Die Funktion der Pericardialdrüsen der Insekten. *Verh. dtsch. Zool.* 1949, 169–73.

PFLUGFELDER, O. (1952). *Entwicklungsphysiologie der Insekten.* Leipzig.

PIEPHO, H. (1938*a*). Wachstum und totale Metamorphose an Hautimplantaten bei der Wachsmotte *Galleria mellonella* L. *Biol. Zbl.* **58**, 356–66.

PIEPHO, H. (1938*b*). Über die Auslösung der Raupenhäutung, Verpuppung und Imaginalentwicklung an Hautimplantaten von Schmetterlingen. *Biol. Zbl.* **58**, 481–95.

PIEPHO, H. (1938*c*). Nicht-artspezifische Metamorphosehormone bei Schmetterlingen. *Naturwissenschaften*, **26**, 383.

PIEPHO, H. (1938*d*). Über die experimentelle Auslösbarkeit überzähliger

Häutungen und vorzeitiger Verpuppung am Hautstücken bei Kleinschmetterlingen. *Naturwissenschaften*, **26**, 841–2.

Piepho, H. (1939*a*). Raupenhäutungen bereits verpuppter Hautstücke bei der Wachsmotte *Galleria mellonella* L. *Naturwissenschaften*, **27**, 301–2.

Piepho, H. (1939*b*). Über den Determinationszustand der Vorpuppenhypodermis bei der Wachsmotte *Galleria mellonella* L. *Biol. Zbl.* **59**, 314–26.

Piepho, H. (1942). Untersuchungen zur Entwicklungsphysiologie der Insektenmetamorphose. Über die Puppenhäutung der Wachsmotte *Galleria mellonella* L. *Roux Arch. EntwMech. Org.* **141**, 500–83.

Piepho, H. (1943). Wirkstoffe in der Metamorphose von Schmetterlingen und anderen Insekten. *Naturwissenschaften*, **31**, 329–35.

Piepho, H. (1950*a*). Hormonale Grundlagen der Spinntätigkeit bei Schmetterlingsraupen. *Z. Tierpsychol.* **7**, 424–34.

Piepho, H. (1950*b*). Über die Hemmung der Falterhäutung durch Corpora allata. Untersuchungen an der Wachsmotte *Galleria mellonella* L. *Biol. Zbl.* **69**, 261–71.

Piepho, H. (1950*c*). Über das Ausmass der Artunspezifität von Metamorphosehormonen bei Insekten. *Biol. Zbl.* **69**, 1–10.

Piepho, H. (1951). Über die Lenkung der Insektenmetamorphose durch Hormone. *Verh. dtsch. zool. Ges.* (Wilhelmshaven), 1951, 62–75.

Piepho, H. and Meyer, H. (1951). Reaktionen der Schmetterlingshaut auf Häutungshormone. *Biol. Zbl.* **70**, 252–60.

Plagge, E. (1938). Weitere Untersuchungen über das Verpuppungshormon bei Schmetterlingen. *Biol. Zbl.* **58**, 1–12.

Plotnikov, W. (1904). Über die Häutung und über einige Elemente der Haut bei den Insekten. *Z. wiss. Zool.* **76**, 333–66.

Poisson, R. and Sellier, R. (1947). Brachyptérisme et actions endocrines chez *Gryllus campestris* L. *C.R. Acad. Sci., Paris*, **224**, 1074–5.

Possompès, B. (1953). Recherches expérimentales sur le déterminisme de la métamorphose de *Calliphora erythrocephala* Meig. *Arch. Zool. exp. gén.* **89**, 203–364.

Poulson, D. F. (1945). Chromosomal control of embryogenesis in *Drosophila*. *Amer. Nat.* **79**, 340–63.

Poyarkoff, E. (1910). Recherches histologiques sur la métamorphose d'un coléoptère (la Galéruque de l'orme). *Arch. mikr. Anat.* **12**, 333–474.

Poyarkoff, E. (1914). Essai d'une théorie de la nymphe des insectes holométaboles. *Arch. Zool. exp. gén.* **54**, 221–65.

Pruthi, H. S. (1924). Studies in insect metamorphosis. 1. Prothetely in mealworms (*Tenebrio molitor*) and other insects. *Biol. Rev.* **1**, 139–47.

Pryor, M. G. M. (1940). On the hardening of the cuticle of insects. *Proc. Roy. Soc.* B, **128**, 393–407.

Przibram, H. (1919). Tierische Regeneration als Wachstumbeschleunigung. *Arch. EntwMech. Org.* **45**, 1–38.

Przibram, H. and Megušar, F. (1912). Wachstumsmessungen an *Sphodromantis bioculata* Burm. 1. Länge und Masse. *Arch. EntwMech. Org.* **34**, 680–741.

RADTKE, A. (1942). Hemmung der Verpuppung beim Mehlkäfer *Tenebrio molitor* L. *Naturwissenschaften*, **30**, 451–2.

RAHM, U. H. (1952). Die innersekretorsiche Steuerung der postembryonalen Entwicklung von *Sialis lutaria* L. (Megaloptera). *Rev. suisse Zool.* **59**, 173–237.

RAMDOHR, K. A. (1811). *Abhandlung über die Verdauungswerkzeuge der Insecten*, p. 66. Halle.

v. REES, J. (1888). Beiträge zur Kenntniss der inneren Metamorphose von *Musca vomitoria*. *Zool. Jb. (Anat.)*, **3**, 1–34.

REHM, M. (1951). Die zeitliche Folge der Tätigkeitsrhythmen inkretorischer Organe von *Ephestia kühniella* während der Metamorphose und des Imaginallebens. *Roux Arch. EntwMech. Organ.* **145**, 205–48.

REITH, F. (1925). Die Entwicklung des Musca-Eies nach Ausschaltung verschiedener Eibereiche. *Z. wiss. Zool.* **126**, 181–238.

RISLER, H. (1950). Kernvolumenänderungen in der Larvenentwicklung von *Ptychopoda seriata* Schrk. *Biol. Zbl.* **69**, 11–28.

SACKTOR, B. (1951). Some aspects of respiratory metabolism during metamorphosis of normal and DDT-resistant houseflies, *Musca domestica* L. *Biol. Bull., Woods Hole*, **100**, 229–43.

SALT, G. (1927). The effects of stylopization on aculeate Hymenoptera. *J. Exp. Zool.* **48**, 223–331.

SALT, G. (1931). A further study of the effects of stylopization on wasps. *J. Exp. Zool.* **59**, 133–66.

SALT, G. (1937). The egg-parasite of *Sialis lutaria*: a study of the influence of the host upon a dimorphic parasite. *Parasitology*, **29**, 539–53.

SALT, G. (1938). Further notes on *Trichogramma semblidis*. *Parasitology*, **30**, 511–22.

SALT, G. (1941). The effects of hosts upon their insect parasites. *Biol. Rev.* **16**, 239–64.

SALT, G. (1952). Trimorphism in the ichneumonid parasite *Gelis*. *Quart. J. Micr. Sci.* **93**, 453–74.

SALT, R. W. (1947). Some effects of temperature on the production and elimination of diapause in the wheat-stem sawfly, *Cephus cinctus* Nort. *Canad. J. Res.* D, **25**, 66–8.

SCHALLER, F. (1952). Effets d'une ligature postcéphalique sur le développement de larves agées d'*Apis mellifica* L. *Bull. Soc. zool. Fr.* **77**, 195–204.

SCHARRER, B. (1945). Experimental tumours after nerve section in an insect. *Proc. Soc. Exp. Biol., N.Y.*, **60**, 184–89.

SCHARRER, B. (1946*a*). The role of the corpora allata in the development of *Leucophaea maderae* (Orthoptera). *Endocrinology*, **38**, 35–45.

SCHARRER, B. (1946*b*). The relationship between corpora allata and reproductive organs in adult *Leucophaea maderae* (Orthoptera). *Endocrinology*, **38**, 46–55.

SCHARRER, (1948). The prothoracic glands of *Leucophaea maderae* (Orthoptera). *Biol. Bull., Woods Hole*, **95**, 186–98.

SCHARRER, B. (1952*a*). Neurosecretion. XI. The effects of nerve section on the intercerebralis-cardiacum-allatum system of the insect *Leucophaea maderae*. *Biol. Bull., Woods Hole*, **102**, 261–72.

SCHARRER, B. (1952*b*). Über neuroendokrine Vorgänge bei Insekten. *Pflüg. Arch. ges. Physiol.* **255**, 154–63.

SCHARRER, B. (1953). Comparative physiology of invertebrate endocrines. *Ann. Rev. Physiol.* **15**, 457–72.

SCHARRER, E. and SCHARRER, B. (1945). Neurosecretion. *Physiol. Rev.* **25**, 171–81.

SCHINDLER, A. K. (1902). Die Metamorphose der Insekten. *Z. Naturw.* **75**, 349–56.

SCHMIDT, C. (1845). *Zur vergleichenden Physiologie der wirbellosen Thiere.* Braunschweig.

SCHMIDT, E. L. and WILLIAMS, C. M. (1953). Physiology of insect diapause. V. Assay of the growth and differentiation hormone of Lepidoptera by the method of tissue culture. *Biol. Bull., Woods Hole*, **105**, 174–187.

SCHMIEDER, R. G. (1939). The significance of the two types of larvae in *Sphecophaga burra* (Cresson) and the factors conditioning them. (Hym: Ichneumonidae). *Ent. News*, **50**, 125–31.

SCHMIEDER, R. G. (1942). The control of metamorphosis in Hymenoptera. *Anat. Rec.* **84**, 514.

SCHNEIDERMAN, H. A., FEDER, N. and KETCHELL, M. (1951). The cytochrome system in relation to *in vitro* spermatogenesis in the Cecropia silkworm. *Anat. Rec.* **111**, no. 164.

SCHNEIDERMAN, H. A., KETCHEL, M. and WILLIAMS, C. M. (1953). The physiology of insect diapause. VI. Effects of temperature, oxygen tension, and metabolic inhibitors on *in vitro* spermatogenesis in the Cecropia silkworm. *Biol. Bull., Woods Hole*, **105**, 188-99.

SCHNEIDERMAN, A. and WILLIAMS, C. M. (1952). The terminal oxidases in diapausing and non-diapausing insects. *Anat. Rec.* **113**, no. 79.

SCHNEIDERMAN, A. and WILLIAMS, C. M. (1953). The physiology of insect diapause. VII. The respiratory metabolism of the Cecropia silkworm during diapause and development. *Biol. Bull., Woods Hole*, **105**, 320–34.

SCHWAN, H. (1940). Beitrag zur Kenntnis der Atmung holometaboler Insekten während der Metamorphose. *Ark. Zool.* **32**, 1–15.

SEAMANS, L. and WOODRUFF, L. C. (1939). Some factors influencing the number of moults of the German roach. *J. Kansas Ent. Soc.* **12**, 73–6.

SEIDEL, F. (1936). Entwicklungsphysiologie des Insekten-Keims. *Verh. dtsch. zool. Ges.* **38**, 291–336.

SEILER, J. (1937). Ergebnisse aus der Kreuzung parthenogenetischer und zweigeschlechtlicher Schmetterlinge. 5. Die *Solenobia*-Intersexe und die Deutungen des Phänomens der Intersexualität. *Rev. suisse Zool.* **44**, 283–307.

SEILER, J. (1951). Analyse des intersexen Fühlers von *Solenobia triquetrella* (Psychidae, Lepid.). *Rev. suisse Zool.* **58**, 489–95.

SELLIER, R. (1946). Transplantation et mue induite d'appendices d'adultes chez *Acheta campestris* L. *C.R. Soc. Biol., Paris*, **140**, 965–6.

References

SELLIER, R. (1949). Diapause larvaire et macroptérisme chez *Gryllus campestris* (Orth.). *C.R. Acad. Sci., Paris,* **228**, 2055–6.

SELLIER, R. (1951). La glande prothoracique des gryllides. *Arch. Zool. exp. gén.* **88**, 61–72.

SHAPPIRIO, D. G. and WILLIAMS, C. M. (1952). Spectroscopic studies of the cytochrome system of the Cecropia silkworm at the temperature of liquid nitrogen. *Anat. Rec.* **113**, no. 78.

SHAPPIRIO, D. G. and WILLIAMS, C. M. (1953). Cytochrome *c* in individual tissues of the Cecropia silkworm. *Anat. Rec.* **117**, 542–3.

SNODGRASS, R. E. (1953). The metamorphosis of a fly's head. *Smithson. Misc. Coll.* **122**, no. 3.

SNODGRASS, R. E. (1954). Insect metamorphosis. *Smithson. Misc. Coll.* **122**, no. 9.

STAUDINGER, F. (1930). Heteromorphosen an Stigmen und anderen Gebilden bei *Carausius* (*Dixippus*) *morosus* Brunner. *Roux Arch. EntwMech. Organ.* **122**, 316–78.

STERN, C. (1941). The growth of testes in *Drosophila*. II. The nature of interspecific differences. *J. Exp. Zool.* **87**, 159–80.

STOSSBERG, M. (1938). Die Zellvorgänge bei der Entwicklung der Flügelschuppen von *Ephestia kühniella*. *Z. Morph. Ökol. Tiere,* **34**, 173–206.

STRICKLAND, E. H. (1911). Some parasites of *Simulium* larvae and their effects on the development of the host. *Biol. Bull., Woods Hole,* **21**, 302–38.

STUTINSKY, F. (1952). Étude du complexe rétro-cérébral de quelques insectes avec l'hématoxyline chromique. *Bull. Soc. zool. Fr.* **77**, 61–7.

SUSTER, P. M. (1933). Beinregeneration nach Ganglienexstirpation bei *Sphodromantis bioculata* Burm. *Zool. Jb.* (*Physiol.*), **53**, 49–66.

SWAMMERDAN, J. (1758). *The Book of Nature; or the History of Insects.* London.

SZTERN, H. (1914). Wachstumsmessungen an *Sphodromantis bioculata* Burm. II. Länge, Breite und Höhe. *Arch. EntwMech. Org.* **40**, 429–95.

TEISSIER, G. (1936). La loi de Dyar et la croissance des Arthropodes. *Libre Jubilaire E.L. Bouvier*, pp. 335–42.

TELFER, W. H. and WILLIAMS, C. M. (1952). The relation of the blood proteins to egg formation in the Cecropia silkworm. *Anat. Rec.* **113**, no. 82.

THOMPSON, W. R. (1923). Sur le déterminisme de l'aptérisme chez un Ichneumonide parasite (*Pezomachus sericeus* Först.). *Bull. Soc. ent. Fr.*, pp. 40–2.

THOMSEN, E. (1942). An experimental and anatomical study of the corpus allatum in the blow-fly *Calliphora erythrocephala* Meig. *Vidensk. Medd. dansk. naturh. Foren. Kbh.* **106**, 319–405.

THOMSEN, E. (1949). Influence of the corpus allatum on the oxygen consumption of adult *Calliphora erythrocephala* Meig. *J. Exp. Biol.* **26**, 137–49.

THOMSEN, E. (1952). Functional significance of the neurosecretory brain cells and the corpus cardiacum in the female blowfly *Calliphora erythrocephala* Meig. *J. Exp. Biol.* **29**, 137–72.

THOMSEN, E. (1954). Studies on the transport of neurosecretory material in *Calliphora erythrocephala* by means of ligaturing experiments. *J. Exp. Biol.* **31**. (In the press.)

THOMSEN, M. (1951). Weismann's ring and related organs in larvae of Diptera. *Biol. Skr.* **6**, 1–32.

TIEGS, O. W. (1922). Researches on the insect metamorphosis. *Trans. R. Soc. S. Aust.* **46**, 319–527.

TOYAMA, K. (1902). Contributions to the study of silkworms. 1. On the embryology of the silkworm. *Bull. Coll. Agric. Japan*, **5**, 73–117.

TRAGER, W. (1937). Cell size in relation to the growth and metamorphosis of the mosquito, *Aedes aegypti*. *J. Exp. Zool.* **76**, 467–89.

VERSON, E. and BISSON, E. (1891). Cellule glandulari ipostigmatiche nel *Bombyx mori*. *Bull. Soc. Ent. Ital.* **23**, 3–20.

VIALLANES, H. (1882). Recherches sur l'histologie des insectes. *Ann. Sci. nat.* (*Zool.*), **14**, 1–348.

VILLEE, C. A. (1942). The phenomenon of homœosis. *Amer. Nat.* **76**, 494–506.

VILLEE, C. A. (1943). Phenogenetic studies of the homœotic mutants of *Drosophila melanogaster*. (i) The effects of temperature on the expression of aristopedia. *J. Exp. Zool.* **93**, 75–98.

VILLEE, C. A. (1944). Phenogenetic studies of the homœotic mutants of *Drosophila melanogaster*. (ii) The effects of temperature on the expression of proboscipedia. *J. Exp. Zool.* **96**, 85–102.

VOGT, M. (1940). Zur Ursache der unterschiedlichen gonadotropen Wirkung der Ringdrüse von *Drosophila funebris* und *Drosophila melanogaster*. *Roux Arch. EntwMech. Organ.* **140**, 525–46.

VOGT, M. (1941). Weiteres zur Frage der Artspezifität gonadotroper Hormone. Untersuchungen an *Drosophila*-Arten. *Roux Arch. EntwMech. Organ.* **141**, 424–54.

VOGT, M. (1942*a*). Induktion von Metamorphoseprozessen durch implantierte Ringdrüsen bei *Drosophila*. *Roux Arch. EntwMech. Organ.* **142**, 129–82.

VOGT, M. (1942*b*). Die 'Puparisiering' als Ringdrüsenwirkung. *Biol. Zbl.* **62**, 149–54.

VOGT, M. (1943*a*). Zur Kenntnis des larvalen und pupalen Corpus allatum von *Calliphora*. *Biol. Zbl.* **63**, 56–71.

VOGT, M. (1943*b*). Zur Produktion und Bedeutung metamorphosefördernder Hormone während der Larvenentwicklung von *Drosophila*. *Biol. Zbl.* **63**, 395–446.

VOGT, M. (1943*c*). Zur Produktion gonadotropen Hormones durch Ringdrüsen des ersten Larvenstadium bei *Drosophila*. *Biol. Zbl.* **63**, 467–70.

VOGT, M. (1946*a*). Inhibitory effects of the corpora cardiaca and of the corpus allatum in *Drosophila*. *Nature, Lond.*, **157**, 512.

VOGT, M. (1946*b*). Zur labilen Determination der Imaginalscheiben von *Drosophila*: II. Die Umwandlung präsumptiven Fühlergewebes in Beingewebe. *Biol. Zbl.* **65**, 238–54.

VOGT, M. (1946*c*). Zur labilen Determination der Imaginalscheiben von *Drosophila*: IV. Die Umwandlung präsumptiven Rüsselgewebes im Bein oder Fühlergewebe. *Z. Naturf.* **1**, 469–75.

VOGT, M. (1947). Beeinflussung der Antennendifferenzierung durch Colchicin bei der *Drosophila* mutante *Aristopedia*. *Experientia*, **3**, 156.

WADDINGTON, C. H. (1940*a*). Genes as evocators in development. *Growth Supplement*, pp. 37–44.

WADDINGTON, C. H. (1940*b*). *Organisers and Genes*. Cambridge University Press.

WADDINGTON, C. H. (1942). Growth and determination in the development of *Drosophila*. *Nature, Lond.*, **149**, 264–5.

WAGNER, G. (1951). Das Wachstum der Epidermiskerne während der Larvenentwicklung von *Calliphora erythrocephala* Meig. *Z. Naturf.* **6** b, 86–90.

WAY, M. J. and HOPKINS, B. A. (1950). The influence of photoperiod and temperature on the induction of diapause in *Diataraxia oleracea* L. (Lepidoptera). *J. Exp. Biol.* **27**, 365–76.

WEBER, H. (1952). Morphologie, Histologie und Entwicklungsgeschichte der Articulaten. *Fortschr. Zool.*, N.F., **9**, 18–231.

WEISMANN, A. (1864). Die nachembryonale Entwicklung der Musciden nach Beobachtungen an *Musca vomitoria* und *Sarcophaga carnaria*. *Z. wiss. Zool.* **14**, 187–336.

WEISS, P. (1940). The problem of cell individuality in development. *Amer. Nat.* **74**, 34–46.

WEISS, P. (1950). Perspectives in the field of morphogenesis. *Quart. Rev. Biol.* **25**, 177–98.

WEISS, P. (1953). The cellular basis of differentiation. *J. Embryol. Exp. Morph.* **1**, 181–212.

WELLS, M. J. (1954). The thoracic glands of Hemiptera Heteroptera. *Quart. J. Micr. Sci.* (In the press.)

WEYER, F. (1928). Untersuchungen über die Keimdrüsen bei Hymenopterenarbeiterinnen. *Z. wiss. Zool.* **131**, 345–501.

WEYER, F. (1935). Über drüsenartige Nervenzellen im Gehirn der Honigbiene *Apis mellifica* L. *Zool. Anz.* **112**, 137–141.

WEYER, F. (1936). Regenerationsvorgänge am Mitteldarm der Insekten. *Verh. dtsch. zool. Ges.*, pp. 157–63.

WIEDBRAUCK, H. (1953). Wiederholung der Metamorphose von Schmetterlingshaut. Versuche an der Wachsmotte *Galleria mellonella* L. *Biol. Zbl.* **72**, 530–62.

WIGGLESWORTH, V. B. (1933). The physiology of the cuticle and of ecdysis in *Rhodnius prolixus* (Triatomidae, Hemiptera); with special reference to the function of the oenocytes and of the dermal glands. *Quart. J. Micr. Sci.* **76**, 269–318.

WIGGLESWORTH, V. B. (1934). The physiology of ecdysis in *Rhodnius prolixus* (Hemiptera). II. Factors controlling moulting and 'metamorphosis'. *Quart. J. Micr. Sci.* **77**, 191–222.

WIGGLESWORTH, V. B. (1936). The function of the corpus allatum in the growth and reproduction of *Rhodnius prolixus* (Hemiptera). *Quart. J. Micr. Sci.* **79**, 91–121.

WIGGLESWORTH, V. B. (1937). Wound healing in an insect (*Rhodnius prolixus* Hemiptera). *J. Exp. Biol.* **14**, 364–81.

WIGGLESWORTH, V. B. (1938). The absorption of fluid from the tracheal system of mosquito larvae at hatching and moulting. *J. Exp. Biol.* **15**, 248–54.

WIGGLESWORTH, V. B. (1939). Häutung bei Imagines von Wanzen. *Naturwissenschaften*, **27**, 301.

WIGGLESWORTH, V. B. (1940*a*). Local and general factors in the development of 'pattern' in *Rhodnius prolixus* (Hemiptera). *J. Exp. Biol.* **17**, 180–200.

WIGGLESWORTH, (V. B. (1940*b*). The determination of characters at metamorphosis in *Rhodnius prolixus* (Hemiptera). *J. Exp. Biol.* **17**, 201–222.

WIGGLESWORTH, V. B. (1942). The significance of 'Chromatic Droplets' in the growth of insects. *Quart. J. Micr. Sci.* **83**, 141–52.

WIGGLESWORTH, V. B. (1943). The fate of haemoglobin in *Rhodnius prolixus* (Hemiptera) and other blood-sucking arthropods. *Proc. Roy. Soc.* B, **131**, 313–39.

WIGGLESWORTH, V. B. (1945). Growth and form in an insect. *Essays on Growth and Form.* Oxford University Press.

WIGGLESWORTH, V. B. (1947). The epicuticle in an insect *Rhodnius prolixus* (Hemiptera). *Proc. Roy. Soc.* B, 134, 163–81.

WIGGLESWORTH, V. B. (1948*a*). The functions of the corpus allatum in *Rhodnius prolixus* (Hemiptera). *J. Exp. Biol.* **25**, 1–14.

WIGGLESWORTH, V. B. (1948*b*). The structure and deposition of the cuticle in the adult mealworm, *Tenebrio molitor* L. (Coleoptera). *Quart. J. Micr. Sci.* **89**, 197–217.

WIGGLESWORTH, V. B. (1948*c*). The role of the cell in determination. *Symp. Soc. Exp. Biol.* no. II, pp. 1–16.

WIGGLESWORTH, V. B. (1948*d*). The insect cuticle. *Biol. Rev.* **23**, 408–51.

WIGGLESWORTH, V. B. (1949). Insect biochemistry. *Ann. Rev. Biochem.* 595–614.

WIGGLESWORTH, V. B. (1952*a*). The thoracic gland in *Rhodnius prolixus* (Hemiptera) and its role in moulting. *J. Exp. Biol.* **29**, 561–70.

WIGGLESWORTH, V. B. (1952*b*). Hormone balance and the control of metamorphosis in *Rhodnius prolixus* (Hemiptera). *J. Exp. Biol.* **29**, 620–31.

WIGGLESWORTH, V. B. (1953*a*). The origin of sensory neurones in an insect, *Rhodnius prolixus* (Hemiptera). *Quart. J. Micr. Sci.* **94**, 93–112.

WIGGLESWORTH, V. B. (1953*b*). Determination of cell function in an insect. *J. Embryol. Exp. Morph.* **1**, 269–77.

WIGGLESWORTH, V. B. (1954). Unpublished observations.

WILLIAMS, C. M. (1942). The effects of temperature gradients on the pupal-adult transformation of silkworms. *Biol. Bull., Woods Hole*, **82**, 347–55.

WILLIAMS, C. M. (1946). Physiology of insect diapause: the role of the brain in the production and termination of pupal dormancy in the giant silkworm *Platysamia cecropia. Biol. Bull., Woods Hole*, **90**, 234–43.

WILLIAMS, C. M. (1947). Physiology of insect diapause. II. Interaction between the pupal brain and prothoracic glands in the metamorphosis of the giant silkworm, *Platysamia cecropia. Biol. Bull., Woods Hole*, **93**, 89–98.

WILLIAMS, C. M. (1948*a*). Physiology of insect diapause. III. The prothoracic glands in the Cecropia silkworm, with special reference to their significance in embryonic and postembryonic development. *Biol. Bull., Woods Hole*, **94**, 60–5.

WILLIAMS, C. M. (1948*b*). Extrinsic control of morphogenesis as illustrated in the metamorphosis of insects. *Growth Symposium*, **12**, 61–74.

WILLIAMS, C. M. (1949). The prothoracic glands of insects in retrospect and in prospect. *Biol. Bull., Woods Hole*, **97**, 111–14.

WILLIAMS, C. M. (1951). Biochemical mechanisms in insect growth and metamorphosis. *Fed. Proc.* **10**, 546–52.

WILLIAMS, C. M. (1952). Physiology of insect diapause. IV. The brain and prothoracic glands as an endocrine system in the Cecropia silkworm. *Biol. Bull. Woods Hole*, **103**, 120–38.

WILSON, E. O. (1953). The origin and evolution of polymorphism in ants. *Quart. Rev. Biol.* **28**, 136–56.

WOLFE, L. S. (1952). Investigations on the structure and deposition of the larval, pupal and imaginal cuticles of *Calliphora erythrocephala* Meigen. Thesis, Cambridge.

WOLFE, L. S. (1954). The deposition of the third instar larval cuticle of *Calliphora erythrocephala*. *Quart. J. Micr. Sci.* **95**, 49–66.

WOLSKY, A. (1937). Production of local depressions in the development of *Drosophila* pupae. *Nature, Lond.*, **139**, 1069.

WOLSKY, A. (1938). The effect of carbon monoxide on the oxygen consumption of *Drosophila melanogaster* pupae. *J. Exp. Biol.* **15**, 225–34.

WOLSKY, A. (1941). Quantitative changes in the substrate-dehydrogenase system of *Drosophila* pupae during metamorphosis. *Science*, **94**, 48–9.

WRIGHT, S. (1945). Genes as physiological agents. *Amer. Nat.* **79**, 289–303.

YOSII, R. (1944). Über den Determinationszustand der Raupenhaut bei der Mehlmotte *Ephestia kühniella* Zeller. *Biol. Zbl.* 64, 305–15.

ZANDER, E. and BECKER, F. (1925). Die Ausbildung des Geschlechtes bei der Honigbiene. II. *Erlanger Jb. Bienenk.* **3**, 161–246.

INDEX

Index

Printed in the United States
By Bookmasters